Springer Series in Advanced Manufacturing

Series editor

Duc Truong Pham, Birmingham, UK

More information about this series at http://www.springer.com/series/7113

Hermann Kühnle · Günter Bitsch

Foundations & Principles of Distributed Manufacturing

Elements of Manufacturing Networks, Cyber-Physical Production Systems and Smart Automation

Springer

Hermann Kühnle
Manufacturing and Production Systems
Otto-von-Guericke-University
Magdeburg
Germany

Günter Bitsch
Stuttgart
Germany

ISSN 1860-5168 ISSN 2196-1735 (electronic)
Springer Series in Advanced Manufacturing
ISBN 978-3-319-18077-9 ISBN 978-3-319-18078-6 (eBook)
DOI 10.1007/978-3-319-18078-6

Library of Congress Control Number: 2015936676

Springer Cham Heidelberg New York Dordrecht London

Printed on acid-free paper

Springer International Publishing AG Switzerland is part of Springer Science+Business Media (www.springer.com)

Foreword I

Industry in general and the metal and electrical industry in particular are subject to highly dynamic forces which have a direct impact on advances in technology and breakthrough innovation. Industry is on the brink of a radical shift in technology—the areas of development and commonly used concepts have already established themselves as part of this sea-change. Innovations such as radio-frequency identification (RFID) chips, smart controls, smart metering, the Internet of Things and embedded cyber-physical production systems are already household terms and represent new, intelligent capabilities within manufacturing. In addition, rapidly evolving new areas, such as big data and cloud computing offer tremendous opportunity while also posing a wide variety of challenges across business organizations.

Südwestmetall—the Baden-Württemberg Employers' Association of the Metal and Electrical Industry—has made these various new developments and technologies the focus of intensive internal discussions and proposed joint initiatives for pursuing these issues with other industry associations. The Industry 4.0 initiative is an important initiative among several that have emerged across Germany. It is designed to prepare enterprises for future developments in this area. As the concept of distributed manufacturing already indicates, ever-increasing globalization will lead to widely distributed organizational structures as well as decentralized manufacturing principles. Because there is still a lack of clearly formulated general concepts in this new area, Südwestmetall welcomes fundamental work and studies which focus on the fast-growing issues underlying the Industry 4.0 revolution. This book presents a comprehensive theoretical discussion of distributed manufacturing in this context. The theme of the book is valuable for the work of our association because the case studies presented can serve as a guideline for management, raise critical questions as well as present possible approaches for working out solutions.

Companies that want to lead the way as this technological shift unfolds and closely follow developments can thus be sure they have a solid basis for making decisions and successfully deal with all the aspects of this new technological challenge.

Peer-Michael Dick
Director General Südwestmetall
Head Office Stuttgart, Germany

Foreword II

In order to successfully operate a globally active organization such as the Constantia Flexibles Group, it is particularly important not only to network the individual production plants, but also the product requirements, production processes and the financial control data and analyses as a basis for decision making.

Only by doing so can client requirements be met in line with market standards, and a sustainable competitive advantage for the company be maintained.

The required global network of the production process, with fast reactivity and flexibility, demands high levels of competence in the set-up, operation and optimization of the production networks. Rigid regulations and operating guidelines are either insufficient or simply incapable of adapting to increasing demands. The operative intelligence of the individual production units needs to be linked in such a way that a context-dependent optimum is achieved at an overall level.

Forward-looking solution concepts need to be designed in such a way that the technological or client-specific adjustments—e.g. in the networking and communication of intelligent sub-components—can either be integrated step-by-step into existing solutions, or existing solutions extended to include them.

The book concerned provides the theoretical basis necessary for these processes, and also gives practical examples of successful implementation.

We are pleased to be able to take some food for thought from this book for the further development of our own organization, for approaches to solutions for existing and future challenges, and, last but not least, for decision-making support.

Dr. Gerold Riegler
Member of the Executive Board, COO
Constantia Flexibles GmbH, Vienna

Foreword III

We live in a world where everything is connected. The network is no longer peripheral—it is central. And that is opening up possibilities—for enterprises and governments, for individuals and society.

With everything connected, distance no longer sets limits on what we can achieve. Mobility gives people tremendous flexibility and freedom, while the rapidly growing Internet of Things (IoT) brings intelligence to billions of objects. The potential—for growth, speed and agility—is limitless.

By using technology more creatively, organizations can achieve stunning business outcomes, and address four key requirements: deliver a great performance, realize possibilities in the cloud, work in harmony and perform anywhere—any time.

In order to deliver a great performance, organizations need to manage a mix of different network technologies—combined with applications performance management and security services. To realize possibilities in the cloud, they require improved performance and availability of multiple cloud-based applications. They connect to different cloud providers and, at the same time, data in the cloud has to remain secure at all times. For an organization to work in harmony, collaboration must happen at any time, on any device. And in order to perform anywhere, they need to deploy great mobility applications on a vast array of devices. But it is not the device that is important, it is the data. So they need the latest security technology to ensure that both data and devices remain secured at all times, regardless of where or how they are accessed or used.

It is all about orchestrating intelligent hybrid systems with a high level of security, while maximizing the performance, flexibility and control offered by the cloud, collaborating with existing Information and Communications Technology (ICT) service providers. This is what we call the art of connecting.

Companies in the manufacturing and logistics industry still move on unstable ground when investing in ICT and outsourcing decisions. As developers of cutting-edge technologies and providers of the above-mentioned services, we understand their needs for support to find their way in this connected industrial world. Value chain networks and manufacturing processes can benefit enormously from the potential the art of connecting delivers. Within the broad domain of distributed manufacturing, it is important to present, discuss and demonstrate the impact of the latest technological developments on the fundamental settings of manufacturing in general.

Distributed Manufacturing articulates the principles and properties of the basic theory and, at the same time, offers examples to facilitate and speed up decision-making. It fulfils all expectations by:

- Proposing the most important insights into upcoming structures and behaviour of distributed manufacturing and illustrating how the intelligent hybrid network brings coherence, how distributed processes integrate and interact with each other
- Demonstrating how mobile and wireless technologies can change manufacturing set-ups in practice and showing how real-time negation of resources—people, technologies and products—configure and enhance the efficiency of manufacturing
- Consolidating impact and trends of the real-time enterprise and cloud-based services anywhere any time, generating innovative principles of concurrency for manufacturing structures and providing examples of successful implementations, verifying at the same time the theory, principles and fundamentals.

Distributed Manufacturing will provide hints to decision-makers and managers. It will support researchers and university teachers to train engineers. The book contains the right mix of theory and practice for promoting the art of connecting in general, and in manufacturing in particular.

In our connected world, the creative use of technology can deliver stunning business outcomes. IT is increasingly an art as much as a science. The objective is to use the power of communications to make a better world. The book focuses on what has been accomplished in the field of collaboration and distributed manufacturing for the benefit of enterprises and governments, individuals and society.

Juergen Meissner
Stuttgart, Germany
BT Global Services

Keith Sherry
Irving TX, USA

Preface

Lector amice salve

Quid mundus, quae causa Deo, ratioque creandi,

Unde Deo numeri, quae tantae regula moli,

Quid faciat sex circuitus, quo quaelibet orbe

Intervalla cadant, cur tanto Jupiter et Mars,

Orbibus haud primis, interstinguantur hiatu

Hic te Pythagoras docet omnia quinque figuris.

Scilicet exemplo docuit, nos posse renasci,

Bis mille erratis, dum fit Copernicus annis,

Hoc, melior mundi speculator, nominis.

Riemannus et Poincareus possunt continuere,

Etiam Hausdorffus et Mandelbrotus

At tu glandibus inventas noli postponere fruges.

Post Mysterium Cosmographicum

Joannis Keplerus

An emerging universe of networkable objects is about to lead Manufacturing into another fascinating phase of development. As the number of smart devices is already counted in trillions, we actually witness their broad introduction into factories. Manufacturing emerges as a set of loosely coupled, smart, autonomous units, which is restructuring configurations and gravitating towards optimising processes and efficient output. Satellites and triangulating transponders are continuously positioning and localising for best transformation steps that are at the centre of interest rather than fixed technological units. Cyber Physical Production Systems, context-aware equipment and networkable manufacturing units become the key assets for industrial value creation. Their agility is orchestrated via intensive aspect motivated interactions, which occur on well-defined orbits with corresponding granularities.

Distributed Manufacturing establishes efficient versatile constellations of skills, advanced technologies and effective processes. It exceeds the simple revising of processes and functions; it incorporates entirely new ways of producing that completely reshape all manufacturing sectors right from the base of collaboration for all involved persons, machines and resources. Attached Cyber Physical Production spaces, Cloud Manufacturing and Smart Objects will evolve to the point where manufacturing needs little prearrangement; it will just happen. Distributed automation has already anticipated the geometry of manufacturing networks in a nutshell; due to pattern similarities, it is now expanding into manufacturing in total. Digital dimensions, which unify virtual worlds with real worlds via interconnected smart mobile devices, make manufacturing once again profit from a number of novelties in the ways we interact with both, the physical and the digital world. Convergence of technologies keeps adding improvements in equipment abilities, in manufacturing productivity and in quality of work life, so networked intelligent units will make Distributed Manufacturing a common way to produce. This book proposes a first solid theory base by outlining overall principles and by detailing the complex nature of smartness of units in Distributed Manufacturing. Especially, the coalescence of people and machines' smartness, now invading manufacturing, is strongly promoted. Outside disciplines are given more credit and easier access to manufacturing. Interdisciplinary work, properly embedded and put together in the right way, paves smooth ways for more and better research results. The outline highlights key outcomes of a number of IMS and ERA Projects, national and regional activities as well as from cutting-edge industrial implementations of Distributed Manufacturing. The results have been consolidated with global engineering communities involving renowned researchers as well as leading manufacturing, telecommunications and software companies. Our thanks go to many university colleagues as well as to a large number of practitioners in different industries from many regions of the world, especially the pilot companies, for their support and insights into recent developments and up-coming concepts.

Stuttgart
Spring 2015

Hermann Kühnle
Günter Bitsch

Contents

Acronyms

AI	Artificial Intelligence
AML	Automation Markup Language AutomationML
AR	Augmented Reality
BAN	Body Area Networks
BOM	Bill of Materials
BOO	Bill of Operations
CAD	Computer-Aided Design
CAE	Computer-Assisted Engineering
CBDM	Cloud-Based Design and Manufacturing
CEP	Complex Event Processing
CIM	Computer Integrated Manufacturing
CIMOSA	Computer Integrated Manufacturing Open System Architecture
CM	Cloud Manufacturing
CPPS	Cyber Physical Production System
CPS	Cyber Physical System
DAMA	Design Anywhere Manufacture Anywhere
DCPS	Distributed Cyber Physical System
DIN	Deutsches Institut für Normung (German), German Institute for Standardization
DMMM	Distributed Manufacturing Maturity Model D3M
DMU	Digital Mock Up
DR	Dual Reality
DSS	Decision Support System
E2AMM	Extended Enterprise Architecture Maturity Model
EA	Enterprise Architecture
ECS	Engineering Client for automation System
EDA	Event-Driven Architecture
EDP	Electronic Data Processing
EE	Extended Enterprise
EPICS	Experimental Physics and Industrial Control Systems Standard
ERA	European Research Area

ERP	Enterprise Resource Planning
ETSI	European Telecommunications Standards Institute
GDTA	Goal Directed Task Analysis
GERAM	Generalised Enterprise Reference Architecture and Methodology
GIS	Geographic Information System
HaaS	Hardware as a Service
HMS	Holonic Manufacturing Systems
IaaS	Infrastructure as a Service
ICT	Information and Communications Technology
IEC	International Electrotechnical Commission
IEEE	Institute of Electrical and Electronics Engineers
IETF	Internet Engineering Task Force
IFAC	International Federation of Automatic Control
IFEAD	Institute For Enterprise Architecture and Development
IFIP	International Federation for Information Processing
IoT	Internet of Things
IP	Internet Protocol
IPSO	Internet Protocol for Smart Objects
ISO	International Organization for Standardization
KAI	Key Alignment Indicator
KPI	Key Performance Indicator
LOD	Level of Detail
M2M	Machine to Machine
MAR	Multiple Augmented Realities
MAS	Multi-Agent Systems
MES	Manufacturing Execution System
MNs	Manufacturing Networks
MR	Multiple Realities
MRP	Manufacturing Resource(s) Planning
NFC	Near Field Communication
P2P	Person-to-Person
PaaS	Platform as a Service
PABADIS	Plant Automation Based on Distributed Information Systems
PDA	Personal Digital Assistant Devices
PERA	Purdue Enterprise Reference Architecture
PLC	Programmable Logic Controllers
POA	Point-Of-Action
POC	Point-Of-Creation
RCE	Resource Cloud Encapsulation
RCM	Remote Cloud Manufacturing
RFID	Radio-Frequency Identification
RTB	Real-Time Business
RTE	Real-Time Enterprise
SA	Service Agent
SaaS	Software as a Service

SCADA	Supervision, Control and Data Acquisition
SCM	Supply Chain Management
SEI	Software Engineering Institute
SIC	Statistical Inventory Control logic
SLP	Service Location Protocol
SO	Smart Object
SOA	Service Oriented Architecture
SoA	Space of Activity
SOAP	Simple Object Access Protocol
STL	Standard Triangulation Language
UC	Ubiquitous Computing
UDDI	Universal Description Discovery and Integration
UPnP	Universal Plug and Play
VAN	Virtual Automation Network
VDI	Verein Deutscher Ingenieure (German) for Association of German Engineers
VM	Virtual Machine
VR	Virtual Reality
WBAN	Wireless Body Area Networks
WSN	Wireless Sensor Network

Chapter 1
Introduction

Rapid developments in ICT totally reshape manufacturing as all machines, objects and equipment on the shopfloors become smart and online. New technology, able to position, to track, to identify on one hand, also capable to communicate, to act, to negotiate and even to decide on the other seems to gain dominating influence on everything in and around manufacturing. Decentralisation and atomisation of processes, units and procedures as well as their complete virtualisations are in trend. Consequently, all interactions between the virtualisations of manufacturing units work exactly like interactions with the units themselves. These virtualisations may even be driven by third party providers on demand, redefining even well established business models. Context aware equipment, on-line monitoring, real time enterprise, autonomous orders, scalable machine capacity or networkable manufacturing unit will be a terminology to get familiar with on the manufacturing shopfloor and in manufacturing management.

All these shifts are strongly fed by powerful cyber-infrastructure, melting information technology and manufacturing resources, closely connecting persons and enterprises; even the manufacturing environment itself is viewing factories as run by intelligent, quasi autonomous distributed units. It is about the optimization of entire value adding process networks; improvements on all relevant fields of distributed automation and process integration provide the potential to reach, to activate and to link units, directly or remotely. Leading companies have been implementing cloud solutions, tracking systems and cyber physical production in selected pilot areas. The results were overwhelming and other shop floor areas have been integrated instantly as well as a number of logistics' functions and multisite topics. There is no doubt that, at long-term, no manufacturing company can get around these subjects. More irrefutable than ever before, the daunting challenge of rapid and complex, technological-driven change triggers disruptive forces on markets, businesses, and economies. If this vision holds, it could be extended to describe the probable next steps of this evolutionary journey of

H. Kühnle and G. Bitsch, *Foundations & Principles of Distributed Manufacturing*,
Springer Series in Advanced Manufacturing, DOI 10.1007/978-3-319-18078-6_1

cyber-infrastructure-based and -assisted enterprises. Humans and units as well as units and units will communicate or even negotiate with the joined objectives of putting together the most efficient manufacturing processes.

As this is widely accepted to be true, the follow-up structures are not yet clear. Manufacturing research as well as advanced practice should therefore provide for wider theory base as well as clearly outlined rationalities for the encountered phenomena, principles and properties. Manufacturing industries will have to quickly internalise the new fields and build up solid skills on networked manufacturing. The discipline of manufacturing is to incorporate other types of methods and tools, supporting linkage and reconfiguration as well as acquisition of high level plug & produce, plug & participate and concurrent work skills; mastering complexity will be crucial and manufacturing structures will have to be enabled for network play!

Along the way, more challenges appear on the horizon, e.g. in the areas of standardization, security, privacy, education, training, regulation, law and policy, standing in the way of the visions' implementation. This time, the disruptions will cut deeply, as these innovations are fostered by convergence of technologies, which makes novelties appear quickly and throughout large sectors, multiplied by network effects. For companies prepared to exploit these innovation forces, new opportunities will offer unprecedented competitiveness. Effective implementations of smart distributed manufacturing networks will make the difference between a leading market position or just simply following the pack with the resulting risk of stagnation and decline. For these reasons, even politics in several regions of the world has put up initiatives, funding schemes and national programs in order to speed up key developments and enhance understanding for this upcoming manufacturing world.

Summarising all observations, technological breakthroughs and expert discussions, it is certainly not too keen to propagate that Distributed Manufacturing is globally on the way to become the preferred model to produce. The easiness to install geographically dispersed manufacturing facilities all over the globe as well as the intrinsic features of high agility, easy presence on volatile markets, cheaper redundancies for higher reliability and so far unseen productivity and resilience are just some major reasons. These reflections include both, the view of geographically distributing manufacturing resources as well as envisioning integrations of manufacturing resources that are already distributed worldwide.

Distributed manufacturing has installed a number of cyber networks in a nutshell in industrial automation already. Groundbreaking pilots are now ready to be extended for constituting more comprehensive solutions as bases for full-scale theory covering both, people's networking and smart manufacturing units' networking.

1.1 Intentions of the Book

The book contributes to a solid base for further work on the rapidly informatising manufacturing field. It primarily addresses the challenges of mastering locally and geographically dispersed intelligent units and proposes techniques, principles and methods for orchestrating smart distributed manufacturing networks. The outline is developed around a number of outstanding project outcomes, fundamentals and exploration results, drawn especially from the Intelligent Manufacturing Systems and the ERA project contexts, as:

- concepts in theory of manufacturing in general and of distributed automation in particular
- properties, key functions and granularity of smart manufacturing units
- existing information models in manufacturing for integrated solutions and distributed concepts
- principles of concurrency along complexity thinking and experiences in distributed manufacturing
- smart manufacturing units' pioneer implementations with anticipated hybrid man-machine decisions
- cyber smartened manufacturing units, using emulation and virtual machines
- human resources' building and managerial implications

As various different disciplines are to be involved, the book lays solid ground for interdisciplinary work and for taking up worthy outside contributions into manufacturing. By a specific layered generic model set up, contributing disciplines may fully stay within their scientific domains on the one hand; on the other hand, interactions between disciplines are instigated for generating more advanced solutions. At this stage, the most important concern is to more deeply integrate human networks and IT networks via stronger coherence. The authors are certain to anticipate and to instigate more innovations by upcoming convergence of fields, e.g. neurosciences and IT, cognitive science and processes, or nanotechnology and IT, expected next within distributed manufacturing. The book proposes overall guidelines and theory for such set ups. To obtain closed and coherent descriptions of networks, topological spaces are introduced as a base. This space construct reduces down to the essentials on one side; on the other side it is powerful enough to fully capture all relevant aspects of manufacturing networks. Even though the generic models, as proposed, are formal mathematical constructs, the outline manages to display all essentials by verbal and figure abstractions, without needing sophisticated formal notations. Smartly attaching adequate models to the network nodes, representing manufacturing units, literally imposes interpretations of cyber physical production and smart objects, as "attachment of models" is simply another wording for the virtualisation of objects.

Any cyber world is a world of interacting models and model systems. Important manners, how these model systems find together, interact and evolve is demonstrated; selected implementations of distributed manufacturing are outlined. Especially intelligent scheduling on the base of smart units as well as intriguing cloud tracing solutions are pointed out, and their DM maturities are evaluated. It becomes clear: Online connections, iterations and 24/24 h availabilities anywhere any time bring up a concurrent distributed manufacturing world with newly appearing principles and properties. The book maps out fundamental concepts of distributed manufacturing into an extended framework for interactions and virtualisations. The outline clarifies the concepts of network control in general and demonstrates how applications of network concepts to industrial organisations in particular.

The book is organised in five main sections:

- Section 1: understanding and defining the working area coming from the fundamental backgrounds (Chaps. 1 and 2)
- Section 2: conjecture of the theory set up emphasising the network nature of manufacturing and fully covering cyber physical systems options. Theory can inform decisions that must be made and additionally prepares tools and techniques that put structure into organisations and processes (Chaps. 3 and 4).
- Section 3: revisiting and structuring information models in manufacturing according to products, processes and decision support (Chaps. 4 and 6)
- Section 4: application and verification of the framework as well as the model world by giving examples from different branches and various solution areas. The examples display practical implementations, their maturities and managerial implications in leading companies (Chap. 6).
- Section 5: conclusions with far-reaching outlook are given in the final section of the book, rounded up by a index supporting to get deeper into this matter easily (Chap. 6 and Conclusions).

A number of well elaborated figures and tables facilitate reading. As around every new subject, extensive fresh terminology has come up in the field of distributed manufacturing. An acronym list as well as an exhaustive index helps the reader to get quickly into the specific language; moreover, a comprehensive literature body for further reading is supplied.

1.2 Intended Audience

This book is especially designed for researchers and scientists on the manufacturing and manufacturing equipment field, machinetool makers, developers of software systems and tools for industry, industry strategists, managers and policy makers in charge of ICT. Machine and equipment makers should consider all cyber developments as disruptive, eventually exposing even the most advanced manufacturing setups to fundamental reconsideration.

The book is also conceived to be a navigator for planners in industry supporting steps towards smartness and online connectivity, a maturity matrix is suggested and the examples highlight a number of important migration aspects.

Lecturers will find an important source for teaching advanced subjects of manufacturing, such as distributed automation, cloud manufacturing, CPPS and, of course, smart distributed manufacturing; even high level master students in engineering and business may profit from some chapters.

Chapter 2
Description of the Working Field

Computing miniaturization and smart devices change manufacturing. Virtualizations and atomization of resources uncover novel manufacturing principles. Virtual representations of units, processes and resources enforce unaccustomed network properties with strategic impact and irrefutable implications on manufacturing. Eventually, many manufacturing setups might have to be reconsidered in total. Mechanisms from distributed automation, already engaging interacting intelligent manufacturing units and decentralized planning and decision procedures illustrate important effects of this irreversible shift of focus towards collaboration and interoperability. Important ICT developments are telecommunication driven and discussed under different chapters, as Internet of things, Ubiquitous Computing, Cyber Physical Systems, Smart Objects or comparable terminology. The subsequent service approaches for informatising and virtualising manufacturing units and processes propagate a number of powerful and fascinating devices, ready for implementation. Properties, which had been found for distributed automation, now reappear for Manufacturing in total, so the upcoming set ups can be mirrored to the experiences and the findings there.

Distributed Manufacturing, as understood here, is a manufacturing network, whose functionality and performance is independent of the physical distance between the involved systems, units and elements. This includes logical and spatial dispersed units, which cooperate and communicate over processes and networks on order to achieve manufacturing functions (Kuehnle 2010). Evidently, the resulting set of loosely coupled, autonomously acting manufacturing units is subject to principles and modes of complex structures, which are unaccustomed and more general, compared to what is known from automation systems and manufacturing set ups (Kuehnle 2012).

H. Kühnle and G. Bitsch, *Foundations & Principles of Distributed Manufacturing*,
Springer Series in Advanced Manufacturing, DOI 10.1007/978-3-319-18078-6_2

2.1 Relevant ICT Thrusts

Intensive researches on the converging fields of telecommunications, computer science and engineering have brought vast progress. Miniaturised, smart, highly mobile, multi-functional electronic devices with enormous computation power offering ubiquitous, uninterrupted, and embedded capabilities are already in use in daily life (Adelstein et al. 2005). In consequence, ICT will be inevitably shaking up the entire manufacturing environment in a next phase. Some important IT developments, as cyber physical worlds, have already provoked intensive reactions in the industrialised world; not only on company levels but on national and international levels as well (Open China ICT 2013). More brilliant innovations e.g. the machine to machine (M2M) communication or body area networks (BAN) are about to rapidly diffuse with radical and disruptive consequences for manufacturing. Their potential can impossibly be ignored by manufacturing companies, as these ICT achievements have already proven to be extremely advantageous and utmost efficient in various other sectors (healthcare, mobility); undoubtedly they broadly find their ways into the manufacturing world.

Some years ago, an object virtualization method has emerged, known as Cyber-Physical System (CPS, DCPS if distributed) (NIST 2011; Sztipanovits 2007; Wolf 2009; Lee 2008), meaning the integration of computing systems with physical processes and physical environments. Initial motivation behind the notion of CPS is the need to design and produce reliable and sustainable computing systems that work in harmony with their surroundings (Ptolemy 2013). Components are networked at every scale and computing is deeply embedded into every physical component, possibly even into materials (Sztipanovits et al. 2012; Derler et al. 2012). Incorporating cyber physical systems into manufacturing environments has brought up the terminology of Cyber Physical Production Systems (CPPS), which is e.g. strongly propagated in the national funding scheme Industry 4.0 in Germany; some advocates even highlight the CPPS context as a fourth industrial revolution (Monostori 2014). CPPS is expected to break up the monolithic functional automation hierarchy into virtualized distributed networked nodes that are more difficult to handle. Cyber Physical Robotics covering industrial robots, tools, sensors, actuators with individual virtual representations of their abilities, constraints and properties, called Cyber-Physical-Devices (CPD) represent Cyber-Physical-Robot-Cells (CPRC) (Michniewicz and Reinhart 2014). When using CPPS (VDI/VDE 2013), components could adapt themselves automatically to the other components, which inevitably changes the way these CPS-enabled components are to be designed and manufactured. Therefore, manufacturers see reasons for totally rethinking industry and industrial production to take the full advantages of CPPS (VDI/VDE 2013).

Parallel to CPS (USA), Computer scientist had come up with the Internet of Things (IoT) in the context of ERA (EU). IoT technologies are already applied for accessing and for connecting manufacturing resources. CPS and IoT cannot be clearly differentiated since both concepts have been driven forward in parallel, although they have always been closely related (CERP-IoT 2009). The IoT can be considered a part of the future internet and be envisioned as a dynamic global network infrastructure with self-configuring capabilities, where physical and virtual objects have identities, physical attributes and virtual personalities; they use intelligent interfaces and they are online connected via information networks.[1] IoT is of vital interest to manufacturing, where the "thing" may typically be the product itself, the equipment, the transportation means, etc.

Triggered by the option of adding more data to any object, we are witnessing the upcoming of a large IoT, where each object has his proper identity (e.g. Eguchi and Thompson 2011); we shall experience smart worlds full of Smart Objects (SO) (Kortuem et al. 2010). A Smart Object (SO) is an autonomous physical and digital object augmented with sensing, processing, and network capabilities.[2] In contrast to RFID tags, SOs carry chunks of application logic that let them make sense of their local situation and interact with human users; coupled with software agent technology however, RFID can transform everyday objects into smart objects as well (Chan et al. 2012). Thus, smart objects and all the developments around are widely discussed as being of highest relevance for manufacturing, as RFIDs are already applied as solution components since years.

Additionally, pervasive computing has migrated from desktops to micro devices, and embedded computing is increasingly integrated into various kinds of objects. Significant progresses have been made in many domains, such as machine-to-machine (M2M) communications using wireless sensor networks (WSNs), as ZigBee[3] and wireless body area networks (WBAN) (Chen et al. 2011). Key achievements refer to easy communication among computers, embedded processors, smart sensors, smart actuators, and mobile terminal devices, without or with limited human intervention (Wan et al. 2012). The rationale behind M2M communication is to generate more autonomous and intelligent applications by networking and interconnecting machines. According to European Telecommunications

[1]Technologies for realizing IoT devices have already been around for years, and have been standardized by the IETF, starting from the lower layers of the stack and moving up. Today, we have IPv6 as a foundation running over links such as those found in mobile networks (2G, 3G and LTE) as well as low power local area sensor networks such as IEEE 802.15.4/6LoWPAN and EPICS. The implementation can be based on multiple agent languages and platforms (JADE, JADEX, LEAP, MAPS) on heterogeneous computing systems (computers, smartphones, sensor nodes).

[2]In 2008, an open group of companies launched the IPSO Alliance to promote the use of Internet Protocol (IP) in networks of "smart objects" (http://www.ipv6forum.com/index.php.) As different definitions of IoT do currently exist, for manufacturing purposes it is useful to refer to IoT as a loosely coupled and decentralized system of smart objects (SOs), which are autonomous physical/digital objects augmented with sensing/actuating, positioning, processing, and networking capabilities.

[3]ZigBee Home Automation is the industry leading global standard helping to create smarter homes that enhance comfort, convenience, security and energy management for the consumer. It appears to be the technology of choice for world-leading service providers, installers and retailers, http://www.zigbee.org/.

Standards Institute (ETSI), standardization plays an indispensable role in long term development of the M2M technology,[4] too. The five elements' structure, as proposed by ETSI, results in three interlinked domains, formed by an M2M area network and M2M gateway, communication network domain and 3G (Lu et al. 2011). A specific Industrial Internet has been brought up by GE (2012), which refers to the convergence of intelligent devices, intelligent networks and intelligent decisions. Industrial Internet is creating the very foundation, needed to make smart manufacturing possible by bringing together brilliant machines, analytics and scalable software platforms to enable nearly instant person-to-person (P2P), person-to-machine (P2M resp. HMI), and machine-to-machine (M2M) communication (Wan et al. 2013).

Ubiquitous Computing (UC), too, has been upgrading objects to smart objects, which can provide customers with new services that could not have been imagined before, because of the steady connection between the real-world objects and the intelligence of information systems. UC denotes another version of a world of smart objects, i.e. physical items whose physical shape and function are being extended by digital components (Langheinrich et al. 2000). UC cannot be considered a proper technology or classified as new functionality, rather a set of functions, which as a sum creates novel computing (Satyanarayanan 2002). Miniaturizations of computer technology provides for smallest processors and sensors for integration into more and more everyday objects, outdating accustomed computing equipment. People will communicate directly with their clothes, watches or pens, and they communicate with each other and with other people's items (Ferguson 2002), surely amazing features with striking relevance for manufacturing.

Cloud computing is a novel model for enabling ubiquitous on-demand network access to a network of computing resources that can be instantly accessed and released with little effort or third party interaction (NIST 2011). A cloud is a parallel and distributed computing net, composed of a set of interconnected computers, presented as one unified computing resource, made available according to service-level agreements (Buyya et al. 2008). Cloud architecture may provide users with options of configurable, virtualized production networks, based on models like federated factories or software-as-a-service applications (Meier et al. 2010). Virtualisations of resources and fast interconnections open up companies

---

[4]The applications of M2M communications extraordinarily depend on many technologies across multiple industries. The technical standardizations for M2M are proceeding in 3GPP, IEEE, TIA, and ETSI. The ETSI drafting standards for information and communications technologies consider an M2M network as a five-part structure http://www.etsi.org/website/homepage.aspx.

(1) Devices, usually are embedded in a smart device to reply to requests or send data.
(2) Gateway, acts as an entrance to another network. It provides device inter-working and inter-connection.
(3) M2M area network, furnishes connection between all kinds of intelligent devices and gateways.
(4) Communication networks, achieve connections between gateways and applications.
(5) Applications and services pass data through various application services and are used by the specific business-processing engines. Software agents analyze data, take action and report data.

in general, and manufacturing areas in particular, to new services and services' architecture, i.e. cloud hardware-as-a-service (HaaS), cloud software-as-a service (SaaS), cloud platform-as-a-service (PaaS), and cloud infrastructure-as-a-service (IaaS). Virtualized computing resources allow, e.g. combined with RFID, Big Data storage and flow. Cloud ERP and Cloud CRM are already available. Online monitoring and positioning of all products and resources enable tracking and fixing issues in real time, allowing companies to instantly improve all attributes of the manufacturing process. With these technologies, the real-time enterprise (RTE) is approaching.

Moreover, a number of researchers already propagate to specify Cloud Computing into Cloud Manufacturing CM, e.g. (Yang 2010), as a new mode of networked intelligent manufacturing, which is service-oriented and highly efficient. CM pushes the development of the concept of Cloud Computing into manufacturing and may become a networked manufacturing mode with quickest responses to market demand, enhanced competitiveness and facilitated collaborative manufacturing (Zhang et al. 2010). Furthermore, Resource Cloud Encapsulation (RCE) of soft and hard manufacturing resources and resource sharing are projected as services for further resource virtualization in CM (Ming and Chunyang 2013). In IoT and CPS, technologies are already widely in use for accessing and connecting physical manufacturing resources. In RCE, all physical manufacturing resources are seen as fully transferred into logical services; based on complete resource virtualization, RCE technology constructs large-scale virtual manufacturing resource pools that can be directly used for interacting and feedback control of manufacturing. RCE is supposed to largely reduce the coupling between physical resource and manufacturing application by the transferring physical resources into logical resources and virtual CM services with instant utilization, high agility, high security and high reliability. In addition, resource pooling and virtualization enable even more sophisticated solutions under Cloud-Based Design and Manufacturing (CBDM). It is a type of parallel and distributed system consisting of a collection of inter-connected physical and virtualized service pools of design and manufacturing resources (Wu et al. 2012) leading to new perceptions of manufacturing.

All cloud solutions enable to dynamically adapt in order to satisfy unpredictable or unexpected demand; manufacturing clouds support scalability to a certain level, e.g. for manufacturing units, general purpose equipment, and standard components for machining. Given that the cloud is a huge shared service pool of design and manufacturing resources, consumers may easily find some dedicated tools and equipment for specific products in the manufacturing cloud that can satisfy their requirements (Wu et al. 2013).

Public clouds are handled by external providers, and the data of various clients may be mixed in factories, servers or CPUs of the cloud.

Private clouds are internally driven and represent a good choice for companies that need high data protection.

Hybrid clouds combine the models of public and private clouds. They may be the key to acquiring an external supply in scaled form and under demand;

however, these clouds add the complexity of determining how to allocate tasks and processes across these different environments (Macia-Perez et al. 2012).

The sum of the ICT progresses, as outlined above, gives plenty of reasons to revisit key portions of our current manufacturing thinking and to justify revising manufacturing from its basics with the aim of setting up a comprehensive framework, which fully captures distributed manufacturing from a consolidated starting point.

2.2 Revisiting Manufacturing Fundamentals

Since its gradual rise from craftsmanship towards industrial structures, manufacturing has made an amazing pathway through history and time. Especially since the late 80s and the beginning of multiple scientific endeavours, manufacturing has undergone quite a number of fascinating developments, as mass production or mass customisation and with lean manufacturing and fractal factory concepts even revolutionary phases. However, what is to come with distributed manufacturing now has every potential to exceed all imagination. An upcoming smart manufacturing world, fully consisting of intelligent, acting, sensing, negotiating and deciding items and units might reach beyond all conceptual capacity; communicating technology objects and parts will orchestrate fragmented and atomised processes and value chains. Manufacturing industries are highlighted by abilities of perfectly mastering long transformation sequences for efficient value creation by implementing the highest degrees of organisation. Other than everyday use smart objects, as smart phones, smart manufacturing units aim at coherent harmonised process chains in the sense of deliberate sequences of transformation, which self-configure and self-optimise, recognised as additional challenges for CPS (Lee et al. 2015).

Currently, much work is on the way generating concepts on how new developments and achievements from information and communication technology may be synthesised best with classical industrial processes. All efforts to achieve more efficient and more flexible production processes seem to be restricted to approaches for men-machine and machine to machine communication and networking of products and machines on the basis of more or less given processes only. However, considering all impacts on the technology domains concerned as well as their possible melting, the backlash on existing processes and overall business models itself may be breathtaking. Just the logical hierarchy of processes and tasks will remain, whereas the physical hierarchy of technological set ups will steadily dissolve. Highly competitive value chains and successful business models will be jeopardised or even be quickly outdated by new versatile ways of putting processes together and playing the game of in- and outsourcing with partners, service providers and expert communities.

Fundamental changes change fundamentals! Therefore, a most rudimentary description of manufacturing as being

- transformation of material by adding
- energy and
- information

into higher valued output is useful to be recalled for identifying and resetting tipping points, which might be ahead.

A short glance at the field of materials reveals: there have been amazing progresses directly impacting most manufacturing transformation technology, process design and all logic behind. E.g. the introduction of plastics, fibre materials and sophisticated metals are already impressive examples and there is certainly much more to come! Novel manufacturing technologies, as additive manufacturing, allow shifting quite a number of manufacturing activities into the design process of a part or a product. Even very sophisticated one-piece products may be manufactured decentralised and rapidly anywhere and anytime, and even the design may be done using collaborative design tools in a distributed way. Embedding IT (nano-) devices into material and parts will make them smarter, and will once more reshape value chains and redesign business models over time. Again, labour division between suppliers, manufacturers, service providers and customers is reset. In distributed manufacturing, all materials and parts may be IT integrated by pervasive computing, by data information codes, by transponder or by other (nano-) devices. Moreover, all materials and objects may be localised and tracked. DNA code-like coining may include all information about the past and the future of the material, part or product.

For this kind of smart distributed objects, enhanced by pervasive computing, the Internet of things, wireless connections and ubiquitous computing, adequate frameworks are still missing and therefore have to be provided. These frameworks should not only embrace all interactions between the manufacturing technology units, the materials and objects, but also interactions between the materials. E.g. after all respective parts already "know" that they will be joined in the final product, their communication capabilities can easily be used to speed up processes to the make value creation more efficient. Moreover, it should lay ground for identifying principles, methods and tools that establish efficient processes using smart materials.

For putting energy to manufacturing transformations, many additional ways have been paved since the early times of water mills and steam engines. A decisive breakthrough was most certainly the wide use of electrical power. Miniaturisation has enabled the self feeding progress in manufacturing technologies via ever-sophisticating transformations steps bringing up full-scale decentralisation of drives and actuators down to the tiniest subunits. Power storage technologies, as well as future options of energy transfer give novel opportunities to disperse, to decentralise and to distribute energy sources and energy plug-in points. Additionally, more options may be kept in mind, basing on wireless energy transfer, smart energy grid, low consumption and efficiency optimised devices for self aware autonomous actions for a number of manufacturing applications.

Formerly, the use of information for executing originally pre-reflected, pre-cogitated transformation steps or respective tasks, was exclusively seen in the context of human intelligence and peoples knowledge, ICTs have come up to take over more and more steps for information processing and information input. Starting with simple NC machines, ICT has meanwhile grown up to the visibly dominating part in all manufacturing equipment. ICT abilities as well as its restrictions have decisive impact on how manufacturing transformation steps are defined, assigned and executed. More and larger transformation chains are taken over by ICT, entire value chains can be performed in full automation and around the clock, so even growing parts of human flexibility in executing tasks and processes will be taken over by modern manufacturing equipment.

Miniaturised distributed databases again allow more and different kinds of decentralisations, therefore, radical changes come about with a new manufacturing world, consisting of autonomous smart items (cyber physical production units, online smart objects or things). This is true for both, for the latest automation solutions as well as for manually driven processes, as technologies may be integrated to participate in this novel manufacturing world. The most prominent application, the distributed automation technology, sees smart manufacturing operations as being fully identified, monitored and controlled by computing and IP communication by sensors embedded in objects and structures of the physical environment. Moreover, distributed automation already provides powerful logic for coupling of sensors and actuators, via fieldbus, Service-Oriented Architectures for modelling of business processes, encapsulations of methods and applications, integration and invocation of services on independent platforms, and agent systems from artificial intelligence. In order to minimise complications with other networks, decentralised and intelligent components are enabled to run local decision processes on model-based descriptions. In automation, bringing in the information for the transformations steps is traditionally done by cascaded control, while time sliced ERP plans will be given down to be detailed on the MES level and executed by field control and plc devices. This entire set up is generally referred to as the automation pyramid, representing a monolithic construct for data transfer and processing.

Innovations in Distributed Manufacturing have brought in agent technology and multi-agent systems, simply dissolving this automation pyramid into autonomous units on the MES level (Kuehnle and Peschke 2007). What had been achieved was the successful introduction of the JAVA virtual machine on jini base for enabling machine controls to communicate via the Internet. This way of completely emulating controls for providing full internet compatibility by using a markup language had initially been seen critical. Today, this is the common approach; the initiated automation markup language AutomationML has been promoted by the authors to the IEC 62714 standard in 2014. Moreover, the ability to communicate via Internet does not only connect the controls online, it multiplies their functionalities by granting full access to a plethora of services that have come up around the Internet.

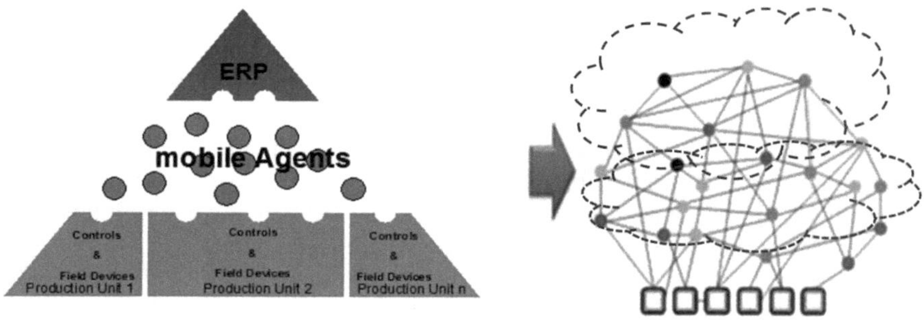

Fig. 2.1 Dissolution of the Pabadis automation pyramid into CPS with hybrid cloud based on fully decentralised automation according to VDI/VDE (2013)

The next wave of innovation, providing smart objects, cyber physical systems and Internet of things will further dissolve the automation pyramid on both, the ERP level and plc or field control levels (Fig. 2.1.). Eventually the structure to cope with will be a large set of cyber physical units and smart objects that are interrelated, online interconnected, self organising, self configuring and self optimising that may optionally be driven by service providers.

This upcoming fully distributed manufacturing world leaves an important issue unsolved: the fallen off hierarchy will have to be replaced by other structures. Smart distributed manufacturing setups, be it within one manufacturing site or in geographically dispersed multiple sites, and surely exhibit all necessary features for configuring manufacturing processes and tools for putting them together. However, their orchestration is profoundly influenced by the designs of production technology and manufacturing equipment, as all components are continuously online and communicate with huge networks of materials, equipment as well as humans. Most recently, we witness a growing and deepening convergence of manufacturing transformations steps with the materials involved and with the information technology units. Each of these domains includes comprehensive technology and systems components' bundles; almost any element may be imagined as being integrated with respective elements of all other domains offering a plethora of powerful additional options for distributed manufacturing. With RFID, smart objects, smart robot modules, micro power cells and smart materials we have only seen the starting points of an amazing take off.

On top of all these powerful innovations, wired and wireless connection and interoperability of all units, materials and energy sources once again give exponential networking combinations of all involved items. Miniaturisation increases abilities and lowers cost at the same time (Moore's law), telecommunication makes fully independent from locations, whilst at the same time allowing to precisely localising items. Traditionally all concentration and modelling efforts had been focused on plans, computerised procedures and numerical abstractions, now the novel identification and localisation capabilities suddenly provide fascinating decision support via online monitoring, tracking, identifying and evaluating real-time data. They enable both; all plans and procedures may be instantly and concurrently matched with manufacturing realities by the full awareness of the whereabouts of all items and the states of all ongoing transformations as well as giving support to decision-making by quick trial and error corrections or iterations, quite different from habitual management thinking. Moreover, all options for Ambient Intelligence (Pallot et al. 2012) as well as Artificial Intelligence become highly available on the shopfloor anywhere and anytime.

The key trajectories as outlined all verify for manufacturing in total, what has been anticipated in a number of distributed automation applications already. Principles and attributes in distributed manufacturing set ups are fundamentally different from what we know so far. One time sequential planning with subsequent command and control is replaced by initial design, iterative adaptations, gradual approximations and evolving perceptions on the base of real-time data. Hierarchy levels exist purely logically, little relevant to organisation or planning horizons; they give decision support background with varying autonomy.

A courageous overall framework for Distributed Manufacturing is called for. Such framework will be provided in the subsequent chapter. Scientific progress here is definitely linked to intense interdisciplinary interaction within solid and consistent overall architecture. As multiple disciplines are involved, solid grounds are laid to scientific interchange instigating theory building and -extension for manufacturing in general as well as by introducing suitable generics in particular. Research work is presented, engaging a number of scientific domains with established theories, as complexity theory, social sciences and network management etc. The authors are convinced that the framework may also capture upcoming patterns of distributed manufacturing, which embrace converging new fields as are already visible, e.g. the nano-technology, neuro-technology or cognition sciences with strong impacts on manufacturing.

References

Adelstein F, Gupta SKS, Richard GG et al (2005) Fundamentals of mobile and pervasive computing. McGraw-Hill, New York

Buyya R, Yeo CS, Venugopal S (2008) Market oriented cloud computing: vision, hype, and reality for delivering it services as computing utilities, CoRR

CERP-IoT (2009) Internet of things strategic research roadmap 15 SEPTEMBER, CERP-IoT, SRA Information Desk European Commission—Information Society and Media DG

Chan C, Chow H, Siu W et al (2012), A multi-agent-based RFID framework for smart-object applications IMECS, Hongkong

Chen M, Leung V, Huang X et al (2011) Recent advances in sensor integration. Int J Sens Netw 9(1):1–2

Derler P, Lee E A, Sangiovanni-Vincentelli A (2012) Modeling cyber-physical systems. Proc IEEE (special issue on CPS, Ptolemy Project) 100(1):13–28

Eguchi A, Thompson C (2011) Smart objects in a virtual world. Int J Comput Inf Syst Ind Manage Appl 3. ISSN 2150-7988

Ferguson GT (2002) Have your objects call my objects. Harvard Bus Rev 80(6):138–144

GE (2012). http://files.gereports.com/wp-content/uploads/2012/11/ge-industrial-internet-vision-paper.pdf
Kortuem G, Kawsar F, Fitton D et al (2010) Smart objects as building blocks for the internet of things. IEEE Internet Comput 14(1):44–51
Kuehnle H (ed) (2010) Distributed manufacturing—paradigms, concepts, solutions and examples. Springer, London
Kuehnle H (2012) Towards production network (PN) theory: contributions from systems of models, concurrent enterprising and distributed manufacturing. Int J E-Bus Dev 2(2):53–61
Kuehnle H, Peschke J (2007) Agent technology and RFID for distributed production simulation and control. In Proceedings of the 14th international EurOMA conference operations management and the new economy. Bilkent University, Ankara, ISBN 978-975-609023-7
Langheinrich M, Mattern F, Römer K (2000) First steps towards an event-based infrastructure for smart things. In: Ubiquitous computing workshop (PACT 2000), Philadelphia
Lee EA (2008) Cyber physical systems: design challenges. In: ISORC'08, Proceedings of the 11th IEEE symposium on object oriented real-time distributed computing. IEEE Computer Society, Washington, DC, USA, pp 363–369
Lee J, Bagheri B, Kao H-A (2015) A cyber-physical systems architecture for industry 4.0-based manufacturing systems. Manuf Lett 3(2015):18–23
Lu R, Li X, Liang X (2011) GRS: the green, reliability, and security of emerging machine to machine communications. IEEE Commun Mag 49(4):28–35
Macia-Perez F, Berna-Martinez J-V, Marcos-Jorquera D et al (2012) A new paradigm: cloud agile manufacturing. Int J Adv Sci Technol 45:47
Meier M, Seidelmann J, Mezgár I (2010) ManuCloud: the next-generation manufacturing as a service environment. Eur Res Consortium Inf Math News 2010:33–34
Michniewicz J, Reinhart G (2014) Cyber-physical robotics—automated analysis, programming and configuration of robot cells based on cyber-physical-systems. Procedia Technol 15(2014):567–576
Ming Z, Chunyang H (2013) Research and realization of resource cloud encapsulation in cloud manufacturing. IJCSI Int J Comput Sci Issues 10(Issue 1 No 3)
Monostori L (2014) Cyber-physical production systems: roots, expectations and R&D challenges. Procedia CIRP 17:9–13
NIST (2011) NIST cloud computing reference architecture, Special Publication 500-292
Open China ICT (2013) European Commission, Document Number D2.1-European Roadmap Survey
Pallot M, Daras P, Richir S et al (2012) 3D-LIVE: live interactions through 3D visual environments. In: Proceedings of the 2012 virtual reality international conference (VRIC'12). ACM, New York, NY, USA
Ptolemy (2013). http://CyberPhysicalSystems.org/CPSConceptMap.xml
Satyanarayanan M (2002) A catalyst for mobile and ubiquitous computing. IEEE Pervasive Comput 1(1):2–5
Sztipanovits J (2007) Composition of cyber-physical systems In: Engineering of computer-based systems, ECBS'07, 14th Annual IEEE international conference and workshops, pp 3–6
Sztipanovits J, Koutsoukos X, Karsai G et al (2012) Toward a science of cyber–physical system integration. In: Proceedings of the IEEE, vol 100, IEEE No. 1
VDI/VDE (2013) Thesen und Handlungsfelder, Cyber-Physical Systems: Chancen und Nutzen aus Sicht der Automation, Verein Deutscher Ingenieure e.V. VDI/VDE-Gesellschaft Mess- und Automatisierungstechnik (GMA), April
Wan J, Chen M, Xia F et al (2013) From machine-to-machine communications towards cyber-physical systems. ComSIS 10(3)
Wan J, Li D, Zou C et al (2012) M2M communications for smart city: an event-based architecture. In: Proceedings of the 12th IEEE international conference on computer and information technology, Chengdu, China, pp 895–900
Wolf W (2009) Cyber-physical systems. Computer 42(3):88–89

Wu D, Greer M-J, Rosen DW et al (2013) Cloud manufacturing: strategic vision and state-of-the-art. J Manuf Syst 5/2013
Wu D, Lane JT, Rosen DW et al (2012) Towards a cloud-based design and manufacturing paradigm: looking backward, looking forward. In: Proceedings of the ASME 2012 international design engineering technical conference and computers and information in engineering conference IDETC/CIE 2012, 12–15 August, Chicago, Illinois, USA
Yang HCh (2010) Cloud manufacturing is a manufacturing service. China Manuf Inf 39(2):22–23
Zhang L, Luo Y, Tao F et al (2010) Key technologies for the construction of manufacturing cloud. Comput Integr Manuf Syst 16(11):2510–2520

Chapter 3
Theory Building Approach

There is an ongoing debate concerning a new theory for manufacturing with focus on collaborative networks (Camarinha-Matos and Afsarmanesh 2008). The call for capturing novel thinking has become prominent since increasing numbers of enterprises are faced with so far unseen challenges to efficiently work in manufacturing networks and distributed structures and thereby operating beyond the consolidated state-of-the-art. Obviously, manufacturing issues have mostly been treated systemic based, along the technological and technical starting points, manufacturing science is built upon. However, this view looses explicative power with the shifts to networks, as the systems thinking hits limits (Dekkers 2013). The scientific community of manufacturing has to recognise that only contributions drawn from other disciplines might ensure more commanding generic concepts, models or theoretical bases for the novel challenges. Among the eligible disciplines that make worthy contributions we may enumerate theory from Complex Adaptive Systems, Decision Sciences, Evolutionary Biology, Game Theory, Organisational Theory and Sociology, alongside more traditional approaches from Network Management. Concepts from Data Exchange, of course, are seen as relevant to structured communication protocols. From all these disciplines and fields contributions have been made to investigate and to describe phenomena of manufacturing networks and related changes that are taking place in industrial entities. By better embedding of practical achievements into theory, key problems in manufacturing networks eventually become more effectively tractable within established research fields. External theories enable managers and stakeholders to reframe issues, to compensate for the unreliability of intuition and common sense, to ascertain belly feelings and to clarify many causal relationships that have impact on firm objectives as well as resource allocation. This phenomenon becomes even more manifest with the latest trends in manufacturing, demanding organisationally and geographically dispersed structures, distributed manufacturing (Kuehnle 2007), demonstrating

H. Kühnle and G. Bitsch, *Foundations & Principles of Distributed Manufacturing*, Springer Series in Advanced Manufacturing, DOI 10.1007/978-3-319-18078-6_3

more loosely coupled entities for industrial networks (Dekkers and Bennett 2009). Fragmentation of manufacturing processes and the developments in information technology have rapidly contributed to higher intelligence and autonomy of factory objects and units.

In manufacturing there have been theory approaches already before that went beyond the systems' thinking (e.g. Forrester 1994). Also, quantitative analysis emerged (e.g. Flynn et al. 1990; Swamidass 1991), inspired by a social science perspective to arrive at theory (e.g. Betrand and Fransoo 2002; Forza 2002) and proved popular for advancing operations management (e.g. Fillipini 1997; Rungtusanatham et al. 2003). However, these developments only led to taking best practice as source for new approaches (e.g. Voss et al. 2002; Davies and Kochhar 2002) (Boyer and Swink 2008) just allowing generalisations yielding limited insight (quantitative research) or solutions for specific circumstances with little attention for wider contingencies (Sousa and Voss 2008), so the domain of manufacturing networks remained entrenched in conceptual approaches with aspirations of (general) systems theories as metatheory solely. However, the developments in manufacturing went on revealing more and more of the network nature as a clear result of the manufacturing processes being in the core and, of course, networks. Even a simple distribution of manufacturing processes may instantly lead to areas beyond state-of-the-art.

More than for any other field, for manufacturing and management science, the dictum: "Nothing is more practical than good theory is" holds true referring to both substantial progress by specific theory as well as for the credibility of managers and practitioners.[1] In this sense, a number of outside theories have already contributed and certainly will further contribute to enhance modes and practices of manufacturing. For instance, the theory for bottleneck management, elsewhere referred to as the Theory-of-Constraints, has shed considerable

[1]The numerous alternative definitions for the term "theory", each of them highlighting specific aspects and emphasising distinct points of view, have all in common that a theory is represented by a set of laws linked by related derivations. For example, the Popperian as well as the Carnapian Schools see theories as sets of statements: scientific theories are general theses and statements that are, as any representation, symbols and systems of characters (Popper 1982). Similar thinking is provided by Sutton and Staw (1995), who regard 'theory' as a set of logically interconnected arguments that tell a story about why certain acts, events, structures and thoughts occur. So, theories do not just ascertain practical insights, theories are considered the bases of all science. And establishing powerful theories is crucial to any scientific progress, but they are also subject to discourse (Foucault 1969). Returning to our line of reasoning, the development of appropriate theories brought considerable progress for manufacturing sciences. A case in point is the broadening material transformations to the total organisational design of manufacturing companies by establishing the Tayloristic thinking, that could later be embedded in the General Systems Theory (Bertalanffy 1973). New ways of modelling, e.g. by interpreting technical transformation as inputs and outputs, allowed deeper insight into the logic of manufacturing organisations and its implications to integration of aspects, decomposition for analysis and appropriate control mechanisms. The resulting thoughts actually are indispensable constituents of all current manufacturing systems' theories.

light on methods for optimisation of inventory and flow of materials in factories. Also, the Theory of Complexity has completely revised the concepts for team structures and for strategy formation in companies, as numerous fractal organisation (Kuehnle 1995) approaches demonstrate. External theories even shifted paradigms in manufacturing; meanwhile the resulting lean concept and the associated complexity thinking, have become constituents of manufacturing science (Kuehnle and Dekkers 2012). In this sense, the domain of Manufacturing Networks and Distributed Manufacturing has been already in a phase of theory-building for some time, pushed by a number of phenomena that have not been explicable within applied and well established theory (e.g. bullwhip effect, instability in turbulent markets). Furthermore, in distributed manufacturing, we increasingly observe research approaches offering frameworks, taxonomies, guidelines, etc., in addition to a number of descriptions drawn from cases and projects as well as approaches built upon allegories and metaphors, e.g. the footprint, the fractal or the holon.

Hence we posit a lack of comprehensive theory for manufacturing networks beyond systems thinking. Tendencies towards interdisciplinarity may eventually be seen as provoked by the specific and rigid profiles of institutionalization and professionalization that science proposes. Viewed from this perspective, interdisciplinarity clearly responds to progressing specialization and knowledge fragmentation as it is widely criticized.[2] Without interdiscipinarity, one might argue that progressing knowledge fragmentation limits innovation capabilities, the usefulness of scientific knowledge or even the relevance of science in general. Scientific progress therefore can only be interdisciplinary and promoted by applied research (Luhmann 1990), in general as well as for Distributed Manufacturing. Particularly, for manufacturing networks, which may be very generally considered as human-governed and systematic combinations of means of technological and conceptual procedures in order to transform inputs into outputs in the sense of marketable products, the phenomena as well have to be described in technological, socio-economical, social, process or strategic perspectives. Consequently, efforts on the domain of Distributed Manufacturing networks should be aimed at intersections with other disciplines and fields, mainly in two directions:

First, the extension of the validity of constructs should be driven forward resulting from the application to aspects from the complex nature of Manufacturing Networks. Such extensions of validity, however, are confined to the prevailing notions and applications; e.g. the common modus operandi of high-frequency adapted traditional planning reacting to the non-foreseeable market movements which can only be coped with by immediate restructurings.

[2]Moreover, there are debates within this context, seriously putting into question scientific specializations and differentiations of disciplinary knowledge as barriers of science and knowledge in total that can be overcome by means of interdisciplinarity (Mittelstraß 1987).

Second, there should, indeed, be efforts of further theorising[3] and to engage disciplines that have already been active in research on Manufacturing Networks more intensively, such as social sciences, information sciences (network's software agents, telecommunication) and management science together with mathematical fields. Comprehensive theory work for manufacturing should preferably take into account distinct disciplines, such as network theories or complexity thinking that already proved to be valuable for addressing distributed manufacturing challenges; more holistic and more comprehensive views are demanded, and open for adaptation of disciplines' borders and for lending from other domains of knowledge.[4]

We argue with that focusing on the intersection of disciplines offers important opportunities to trespass boundaries, to redefine core issues and to foster further theory building. Developing skills at intersections gives rise to issues of legitimacy, paradigm convergence, interdisciplinary communication as well as fresh answers to complex phenomena. There are three modes that researchers can employ with varying levels of impact—ranging from (1) the mere borrowing of concepts, (2) the extension of original theories with the more ambitious redefining boundaries to (3) the transforming of the core of parent fields and disciplines by new domains (Table 3.1).

Because theory building is likely to generate the richest insights, transforms core disciplines by defining and consolidating domains, as Distributed Manufacturing constitutes, our choice has to be for Mode 3, as it offers solid theory building opportunities.

For distributed manufacturing and manufacturing networks a set up would be adequate, when it allows addressing and assigning attributes and indicators for all

[3]In that context, science may be named any intersubjectively verifiable examination of facts, including their systematic descriptions and—if possible—their explications (Carnap 1966). With an identified object of interest as a starting point, any science traditionally strives for understanding and principles in line with specificities of the associated branch of knowledge, also referred to as the accounting scientific discipline within the relevant classification of sciences (Popper 1959). Well-established scientific disciplines have considerable impact on research. The content of theory to be proven seems to strongly depend on presumptions, experiential evidence and ad hoc explanations that constitute scientific progress, however always tightly held together by a dominant paradigm that may as well be referred to as the identity of the accounting discipline. In this perspective, we speak of a pure discipline or of mono-disciplinarity if a certain domain is scientifically permeated with a consistent paradigmatic and theory-rich concept.

[4]Theory building, e.g. (Weick 1995), in general, occurs through distinct procedures, as induction, deduction, comparative analysis or theoretical sampling within a discipline and eventually the formation of further reaching theories (Glaser and Strauss 1967; Suddaby 2006). In such instance, principles of abstraction (Dekkers 2013; Timpf 1999)—classification, aggregation and generalisation—will support the extension of principles and solutions to become underpinning theories. With formations as e.g. Holonic Manufacturing or Soft Artefact, valuable scientific qualities came to light to be theorised, in synthesis with complexity theories and life-cycle approaches. Similarly, skilled conjectures for network manufacturing will cumulate insights into adequate frameworks more swiftly and more coherently through engaging with more widely accepted impactful models embracing strong bodies of (ready for use) knowledge.

Table 3.1 Comparison of the three modes of theory building at the intersection of disciplines (Zahra and Newey 2009)

Dimension	Modes		
	1	2	3
Focus	Importation of theory from other fields	Intersect on of one field/ discipline and another	Using intersection as a new domain to develop new theory
Objective	Enrichment and extension	Boundary definition	Domain definition
Critical kills	Empirical	Synthetic integration	Theory building cross disciplinary communication
Use	Widespread	Frequent	Rare

relevant manufacturing objects. Moreover, full unit descriptions, capabilities' models and objectives and bundles should be assignable, and communication and decision capabilities should be incorporated. Following the principles of engineering, any set up may be accepted as theory, if it addresses most problems, and if it is currently describing and solving problems at the highest rate (i.e. Matheson 2014). In this respect, we go along with (Mintzberg 2005), who advocates that theory can be seen along a continuum, from lists (categories), to typologies (comprehensive lists), to impressions of relationships among factors, to causations between and patterns among these relationships and to fully explanatory models.

Consolidation work on scientific theory may be achieved by:

1. Improving congruence of observations and predictions,
2. Defining quantitative or phenomenological laws,
3. Outlining master examples for the solution of scientific problems including the incorporation of new discovered phenomena (Kuhn 1962).

A construct may definitely be envisioned as theory, if designs for the description of contexts and a comprehensive frame are melting substantial interdisciplinary contributions tightly together. Since mathematical tools represent a common language, facilitating communication among and between disciplines, mathematical set ups are generally given "natural" superiority in interdisciplinary contexts.

In this sense, the frame for adequate design of distributed manufacturing networks' theory is proposed as a collection of shells around a formal core and generic elements, enclosed by a shell of specific laws, principles and rules respectively. These are embraced by another shell, the shell of models, that may either be newly established or frequently in use within the context. These three shells are viewed as embedded in the real manufacturing world context, which is the practitioners view, and the field for successful models' implementations.

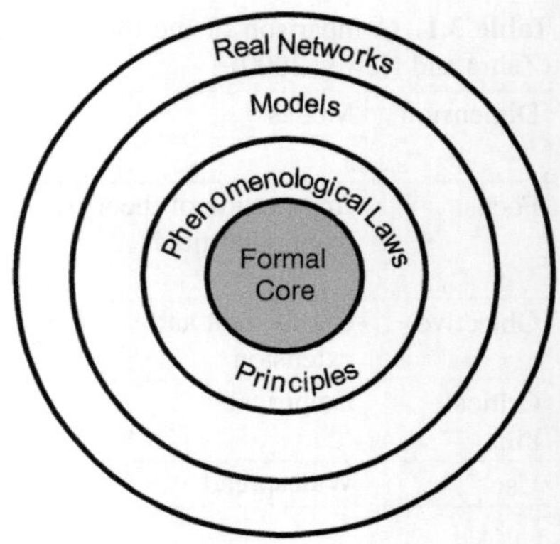

Fig. 3.1 Distributed manufacturing theory design: models, derived from formal core and generics interact with real networks according to guiding principles and phenomenological laws

The conjecture may be visualised best as concentric circles' construct (Fig. 3.1), with

- a centred formal theoretical core and generics,
- the shell of principles and phenomenological laws,
- the models shell and
- an orbit of real world examples.

As networks are, according to their nature, objects of geometry, we talk about abstractions and virtualisation of geometry in the context of manufacturing networks as well. A list of undoubtedly contributing disciplines, as graph theory, network management or data communication/data exchange, verifies the statement. The mathematical theory behind is topology, or more precisely, the theory of manifolds and its specific spaces, ideally support the integration of models and their interactions, via self-similarity (Mandelbrot 1985) compositions and operations. The well established Hausdorff space, a specified topological space (Boto von Querenburg 2001), embraces brilliant backgrounds for interpreting manufacturing networks, as it envisions all network units as nodes "carrying" rich model structures. To any element or unit (point) of a MN, a set of tangent spaces (as subsets of E^n resp. $R^{n)}$ is assigned by mappings, allowing to embed different models e.g. machine geometries, capabilities and KPIs (Fig. 3.2), in manufacturing referred to as the virtualisations of a unit. Moreover, these virtualisations may encapsulate even richer properties and attributes.

The Hausdorff space proves rich enough to capture a vast majority of configurations occurring in distributed manufacturing and manufacturing networks. These configurations may be exhibited by connectors, relations and interactions between the "attached" models, assigned to the manufacturing networks nodes. If advantageous or simplifying, the manufacturing networks, its attributes and its configurations may arbitrarily "forget" or "remember" attached models allowing perfect

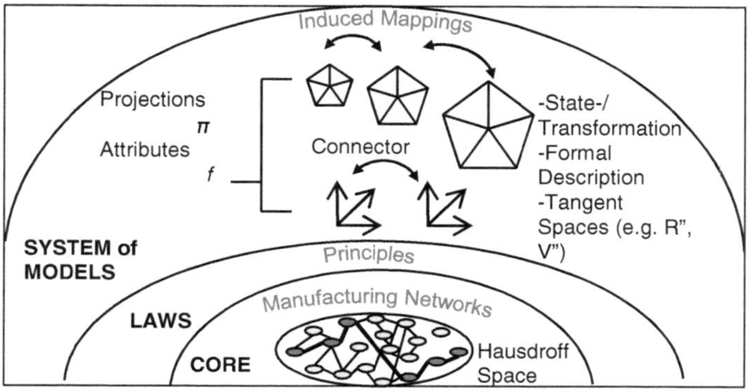

Fig. 3.2 Manufacturing network as hausdorff space with space of activity (Tangent Space) example and units' state observable (projection levels) and connectors

procedures e.g. to capture encapsulations, fold and unfold properties, or on-off modes of behaviour, e.g. self-organisation (Mandelbrot und Evertsz 1991).[5] Configurations may also be modelled by attributes and indicators expressing specific disciplinary views for attached tangent spaces, compiling different views of the network nodes.

Key theory portions are to be represented by elements and structure mappings preserving structural properties while operations as unfold/fold or expansion/compression modes are applicable. Introducing the self-similarity principle (Kuehnle 2009), fractal structures may capture the imperatives of homomorphism of the mappings, resulting in valuable hints for harmonisation requirements concerning interacting units on and between all levels of detail, an important point for further standardisations. More formal details are given in the Annex of this book.

For network any descriptions there is this need for formalising connectivity; moreover, the core models should provide easy access for multiple disciplines to be employed in multiple areas of DM. These requirements are easily met by the introduction of generic models, open enough to gather different disciplines and approaches and at the same time instigating disciplinary interaction by a closed unified description.

[5]Without going into further details, it can be stated that topology has had an impact manufacturing networks' research already. By introducing topology, many portions of manufacturing theories (e.g. generic elements, models and principles for social agents as well as software agents' network interactions) were reframed; other elements permit designing novel steadily evolving network decision modes and such set-up facilitates exploiting the networks' characteristics related to cooperative games and partnership for value optimisation. These examples of topology influence stand for quite a number of effects that have been observed around Manufacturing Networks.

References

Bertalanffy LV (1973) General system theory. George Braziller, New York

Bertrand JWM, Fransoo JC (2002) Operations management research methodologies using quantitative modeling. Int J Oper Prod Manage 22(2):241–264

Boto von Querenburg (2001) Mengentheoretische topologie. Springer, Heidelberg

Boyer KK, Swink ML (2008) Empirical elephants—Why multiple methods are essential to quality research in operations and supply chain management. J Oper Manage 26(3):337–348

Camarinha-Matos LM, Afsarmanesh H (2008) Collaborative networks: reference modeling, Springer, Berlin

Carnap R (1966) Philosophical foundations of physics: an introduction to the philosophy of science. Basic Books, New York

Davies AJ, Kochhar AK (2002) Manufacturing best practice and performance studies: a critique. Int J Oper Prod Manage 22(3):289–305

Dekkers R (2013) Applied systems theory. Springer, London

Dekkers R, Bennett D (2009) Industrial networks of the future: review of research and practice. In: Dekkers R (ed) Dispersed manufacturing networks: challenges for research and practice. Springer, Heidelberg, pp 13–34

Filippini R (1997) Operations management research: some reflections on evolution, models and empirical studies in OM. Int J Oper Prod Manage 17(7):655–670

Flynn BB, Sakakibara S, Schroeder RG et al (1990) Empirical research methods in operations management. J Oper Manage 9(2):250–284

Forrester JW (1994) System dynamics, systems thinking, and soft OR. Syst Dyn Rev 10(2–3):245–256

Forza C (2002) Survey research in operations management: a process-based perspective. Int J Oper Prod Manage 22(2):152–194

Foucault M (1969) L'Archéologie du Savoir, Gallimard, Paris

Glaser BG, Strauss AL (1967) The discovery of grounded theory: strategies for qualitative research. Aldine, New York

Kuehnle H (1995) L'entreprise fractale. In: Braesch C, Haurat A (eds) La modélisation systémique en entreprise, Pôle productique Rhône-Alpes, Paris, pp 263–272

Kuehnle H (2007) Post mass production paradigm (PMPP) trajectories. J Manuf Technol Manage, Emerald 18:1022–1037

Kuehnle H (2009) Self-similarity in dispersed manufacturing: a contribution to productions networks control. In: Dekkers (ed) Dispersed manufacturing networks, Springer, Berlin

Kuehnle H, Dekkers R (2012) Some thoughts on interdisciplinarity in collaborative networks' research and manufacturing sciences. J Manuf Technol Manage, Emerald 23:961–975

Kuhn T (1962) The structure of scientific revolutions. University of Chicago Press, Chicago

Luhmann N (1990) Die Wissenschaft der Gesellschaft, Frankfurt

Mandelbrot BB (1985) Self-affine fractals and fractal dimension. Phys Scr 32:257–260

Mandelbrot BB, Evertsz CJG (1991) Multifractality on the harmonic measure on fractal aggregates, and extended self similarity. Phys A 177:386–393

Matheson C (2014) Historicist theories of scientific rationality, First published 12 Aug 1996, substantive revision 15 Aug. Last accessed 02 Apr 2015

Mintzberg H (2005) Developing theory about the development of theory. Oxford University Press, Oxford

Mittelstraß J (1987) Die Stunde der Interdisziplinarität?, Kocka, J. (Ed.) Interdisziplinarität. Praxis – Herausforderung – Ideologie, Frankfurt

Popper K (1959) Logic of Scientific Discovery, London

Popper K (1982) Logik der Forschung (7.Aufl.), Tübingen (first 1934: Logik der Forschung, Springer, Wien)

Rungtusanatham MJ, Choi TY, Hollingworth DG et al (2003) Survey research in operations management: historical analyses. J Oper Manage 21(4):475–488

Sousa R, Voss CA (2008) Contingency research in operations management practices. J Oper Manage 26(6):697–713

Suddaby R (2006) From the editors: what grounded theory is not. Acad Manag J 49:633–642

Sutton RI, Staw BM (1995) What theory is not. Adm Sci Q 40:371–384

Swamidass PM (1991) Empirical science: new frontier in operations management research. Acad Manag Rev 16(4):793–814

Timpf S (1999) Abstraction, levels of details, and hierarchies in map series. In: Freksa C, Mark DM (Eds.), Spatial information theory—cognitive and computational foundations of geographic information science, vol 1661. Springer, London, pp 125–140

Voss C, Tsikriktsis N, Frohlich M (2002) Case research in operations management. Int J Oper Prod Manage 22(2):195–219

Weick KE (1995) What theory is not, theorizing is. Adm Sci Q 40:385–390

Zahra SA, Newey LR (2009) Maximizing the impact of organization science: theory-building at the intersection of disciplines and/or fields. J Manage Stud 46(6):1059–1075

Chapter 4
Core Models, Principles, and Networks' Structuring

With the topology framework, generics are proposed and model worlds recalled. Principles and laws follow, which had been observed in the set ups of distributed automation and concurrent enterprises, now ready to be generalised. All involved models are detailed; for harmonising the models on the respective levels of detail, the models ideally stem from the generics. On one hand the generics should demonstrate fold and unfold properties in a way that units' virtualisations are able to offer correspondent granularities. On the other hand, they should be general enough to be adapted to the respective units' content.

Manufacturing units may be seen as a special kind of smart objects (Kawsar and Nakajima 2009) or specifications of CPS, or as a part of the IoT. Moreover, manufacturing will increasingly appear equipped by devices, featuring sensing, processing, actuating and networking capabilities. Additional computing power widens abilities, as environment-awareness or self-logging and self-reporting, and further augments the units by storing more data about themselves and their activity domains. In consequence, manufacturing units may e.g. make emergence network structures as collaboration results between units. Attracted by incentives, the units float within network configurations, communicate, negotiate, link or breakup on all levels of detail. Manufacturing networks, thus, are orchestrated by self-optimising and self-orienting units following objectives and defined rules when initiating interactions or configuring processes. The governing attractors are derived from business opportunities as inputs for driving, operating and restructuring manufacturing in the sense of versatile collaborative process networks. From the network point of view, some configurations are more favourable than others with respect to the objectives; therefore, continuous monitoring has to evaluate and to take actions for gradual and stepwise decisions on configuration alternatives. Orchestrating a manufacturing network includes all levels of detail; these levels are reached by self-similar breakdown and fractalisation.

H. Kühnle and G. Bitsch, *Foundations & Principles of Distributed Manufacturing*, Springer Series in Advanced Manufacturing, DOI 10.1007/978-3-319-18078-6_4

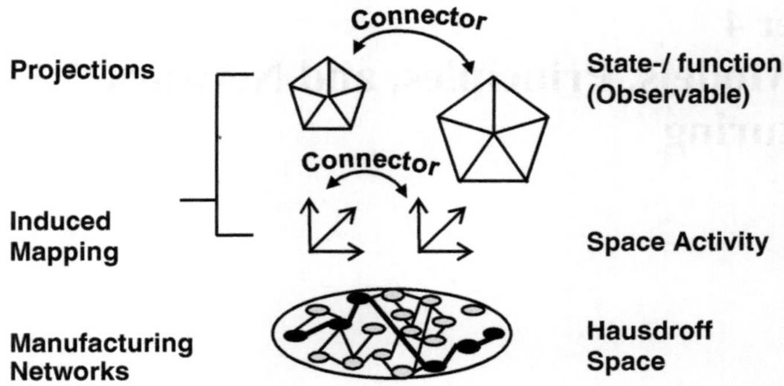

Fig. 4.1 Manufacturing network as hausdorff space with spaces of activity (Tangent Spaces) capturing aspects and levels of detail (morphisms)

Envisioned as such, manufacturing networks appear as levelled networks (Fig. 4.1) composed of smart autonomous manufacturing units mappings, which communicate on all levels of detail.

4.1 Generic Models in the Core

The key abstractions in the core should, in the first place, capture key attributes of nets and units and, second, provide easy access for multiple disciplines to be employed in their respective fields. The models have to represent the units on all levels of detail; their varying configurations may also be represented by indicators and attributes, as key constituents.[1] Any manufacturing setup will derive objectives for its output (products, product programs), for the components and necessary parts, as e.g. described in the bill of materials. These objectives will then induce to make coordinated use of the units and their resources, which include technical functions and operations. Envisioned like this, a) the objective view and b) the resource view are just different "sides" of one and the same unit. Suitable generic modelling must, therefore, include details of both sides as well as strategic and economic issues on all levels.

Concerning bundles of objectives, their case dependant priorities leverage complementing and conflicting issues. For depicting the objectives of manufacturing networks and units, a specific triangle stands for the priorities (Fig. 4.2). Derived from

[1]In the resulting topology, all projections should be homeomorphisms, so utmost compatibility of the models is achieved. For implementations, this postulate already states the need for standards ensuring this networkability. The self-similarity principle perfectly captures the nature of homomorphism of these mappings (Kuehnle 2009) and, therefore, gives valuable hints for harmonisation, concerning corresponding structures on and between all levels of detail.

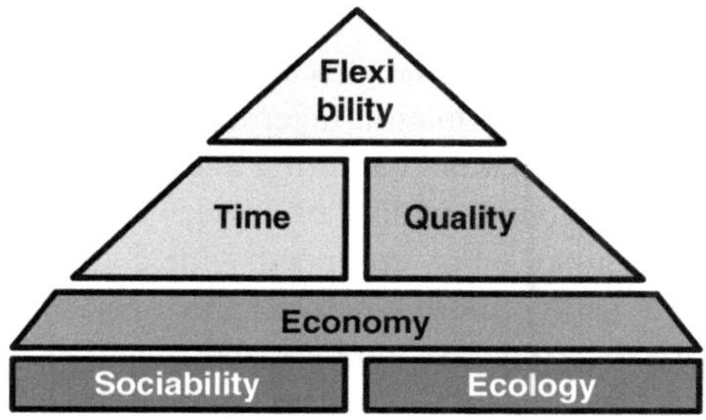

Fig. 4.2 Generic objectives' priority set-up for network and units

strategy, it constitutes the tactical generic pattern to express case specific objectives and indicators. Subsequent fold and unfold operations detail or compress these pattern using KPIs or key alignment indicators KAIs (Piedade et al. 2012) for evaluating network configurations or for matching capabilities. All units' objectives are to be harmonised with the overall network objectives, so consistency procedures, e.g. negotiation consensus, for all levels of the manufacturing network must be provided. The fulfilment of objectives within a constellation does imply an intensive communication between the levels with self-similarity ensuring the intended compatibility patterns on the sub levels as well as between units.

For manufacturing networks, aspect wise full decompositions have already been successfully introduced by the authors as generics for fractalising organisations and fractal enterprises, distinguishing between aspects as information, organisation and processes. The intention and the setups are comparable to e.g. the specification of the CIM/OSA framework and consecutive standards (Kosanke 2006) for seamless communication. Equivalent layer wise approaches for resource co-ordination schemes for manufacturing have also been applied elsewhere (Bartlett and Goshal 2002; Alt and Smits 2007) for enterprise units' networks and in the contect of cyber physical systems principles (Sacala and Miosecu 2014; Lee et al. 2015). Here, a layer wise decomposition is proposed, apt for folding and unfolding generics to support levelled networkability on all layers and exhibits all relevant aspects together and separable, Fig. 4.3. For the core model purposes, these layers have proven to brilliantly represent all relevant items and are suggested here as the subsequent aspect layers (Kuehnle 1995; Kuehnle and Schmelzer 1995):

I. The culture layer envisions the network as a social system and captures the value and thinking pattern within the network, sometimes resp. broken down onto the units (site culture) and sub-units (team culture). Consistent values are prerequisites for any networks' success.
II. The strategy layer describes the way, in which the network deals with competition, the market and the resources. To quantify strategies, networks use objective systems, describing the actions of a network towards markets, economical pressure and technological changes and define objectives.
III. The social-informal layer grasps all human resources and organisation contexts in the network. It includes formal factors, as responsibilities or task assignments, and informal factors that influence and determine social relationships within the network, teaming and communication skills.

IV. The financial layer represents tools and instruments for evaluation of performance and the allocation of value addition across the network.
V. The information layer addresses the total design and handling of the flow of information; capturing and putting out data, backing up interconnections and re-configurability of devices by IT infrastructure. Special concerns in DM are smart units, equipped with computing units; control systems are emulated using different networkable operating systems.
VI. The layer of process and material flow addresses all technical and physical sides of the transformation steps, direct or indirect. Technical function and transformation descriptions as well as logistics and materials' handling are covered.

The layers culture and strategy may be considered as guiding for all other layers, summarised as resource layers. Dependent on the case and the level of detail to be addressed, fold and unfold properties are applicable to meet the corresponding levels of detail for communication between different entities on all layers. The layers also support the syntheses of networks, allowing profound network modelling by highlighting distinct aspects e.g. human centred team concepts or purely ICT driven units. Moreover, layer-wise linking and interconnection of units maintains descriptions of aspects throughout the entire networks on one hand and on all levels of detail (Fig. 4.3) on the other.

Eventually, for any of the layers, all levels of detail are available for single aspect considerations, may it concern an element or the total network.

The composed six layer stack may be used as generic compound in total; it embraces all key constituents according to the requirements, including objectives, resources, and inputs and outputs as displayed. The six layer stack in total as well as objectives and resources individually may be broken down to the element level or folded up to network levels respectively (Fig. 4.4).

As an overall result, all levels of detail are available in parallel for all layers which can be detailed for sub packs or single aspect considerations. Units' interrelations may be handled in the same manner. This way the units appear as composites having exactly the same functional model but are composed by a set of lower

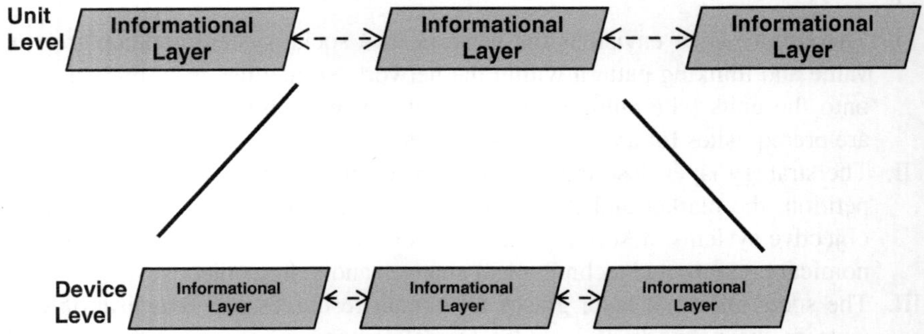

Fig. 4.3 Units' informational layer, replicated on the shopfloor level by break down into subunits

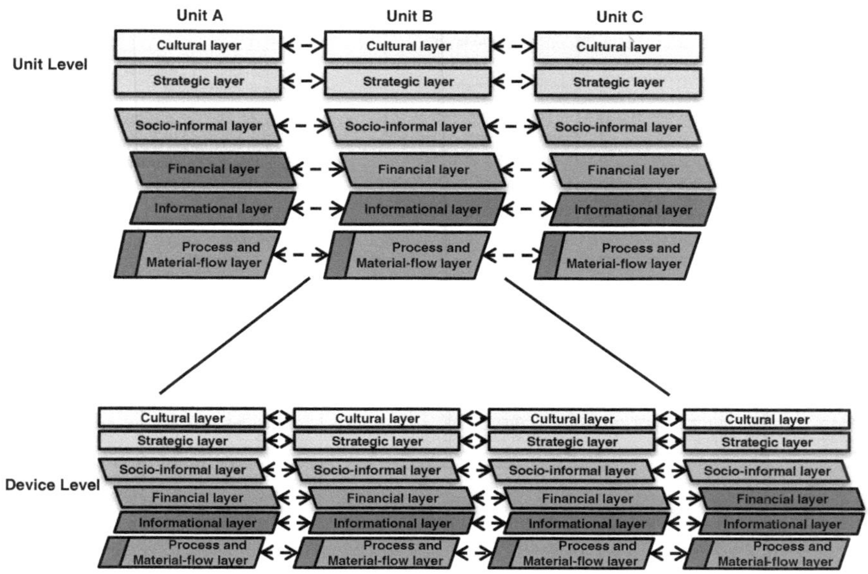

Fig. 4.4 Sequence of units configuring processes, with unit B broken down onto the shopfloor level by all six aspects respectively layers

level subunits on the shopfloor. Up fold aggregation effectively creates an equivalent on the factory level and on the network level.

Moreover, embedding the six layers stack, the generic description opens up to other scientific fields and their concrete contexts in the respective industrial applications. This is not only proposed for a specific granularity but for all levels of detail.

Each layer is ready to interact with disciplines from other layers, is open to embrace worthy contributions from other fields to be eventually incorporated into one unified model, and at the same time layers' contributions remain separable for their respective disciplines (Fig. 4.5). As the most common example, the parallel visualisation of the process layer and the economic layer including KPI's may be hinted at the visual systems design VSD (Rother et al. 1999). Other examples are parallel set ups of the social informal layer and the process layer expressing responsibilities for tasks and transformations. All these features are proposed on all levels of detail and may be synthesised and broken down according to necessities.

For Distributed Manufacturing and the units involved there, primarily the functional dimension is to be addressed. Therefore, orders, process segments and tasks, executable by technology, people or by means of software and software agents[2]

[2]Groundbreaking work on the agent field has been done in international projects and by multinational consortia for establishing standards (AML) that support distributed communication and decision making structures. Results, closest to the areas outlined here, are the set ups of PABADIS (Peschke et al. 2005) and GRACE (2013).

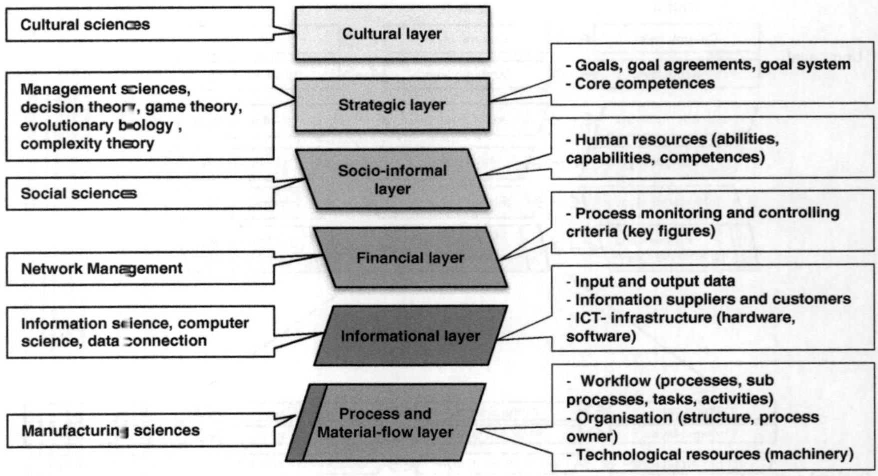

Fig. 4.5 Process and manufacturing unit exhibited by the generic 6 layer description, additionally offering outside contributions from other scientific fields and theories including interactions

negotiate, self configure and self control. For orchestrating such functional units, a simplification is favourable; the six layer aspects may be reduced to a more compact model and depicted as a three-dimensional space. The resulting figure, with the dimension of objectives, the dimension of resources and the restrictions concerning the respective unit, is named Space of Activities (SoAs) (Fig. 4.6). It constitutes the network unit projection of the six layers and the indicators respectively and synthesises the objective layers and the resource layers by more aggregate descriptions. The SoA volume may then be interpreted as the decision space of a unit, which represents the environment a unit can rely upon for task fulfilment and performance outcomes (according to specific unit's performance parameters) as well. A unit's performance indicates (given by respective measuring and monitoring) the unit's behaviour with impact on self-optimizing modes, on turning on, -down or -off a unit's autonomy as well as on dimensioning its total network's intervention. If a unit constantly fails to supply outcome within the space of activity, i.e. to self optimise and to adapt to the challenges of the surroundings, higher network levels may come into take over the units' activities.

The spaces of Activity lay ground for the mechanisms, which gradually adapt manufacturing networks (Ivanov et al. 2006). Most favourable for optimising manufacturing networks seem to be procedures, engaging distributed, simultaneous logic and continuously generating small step progresses (Bennet and Bennet 2004). Series of network units' decisions on all levels will gradually evolve the networks for better meeting the objectives, for economising resources and, thereby, strengthening and enhancing the networks' competitiveness.

The layer stack as outlined may be seen as the generic process segment description. Units contribute into network configurations by negotiation and design of

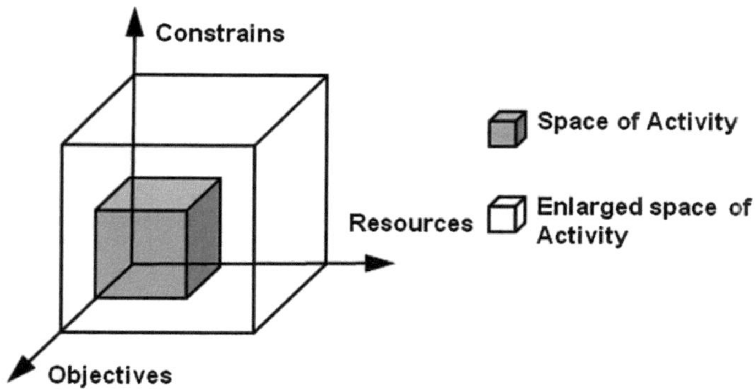

Fig. 4.6 Units' Space of Activity (SoA)—monitoring projections of relevant autonomy

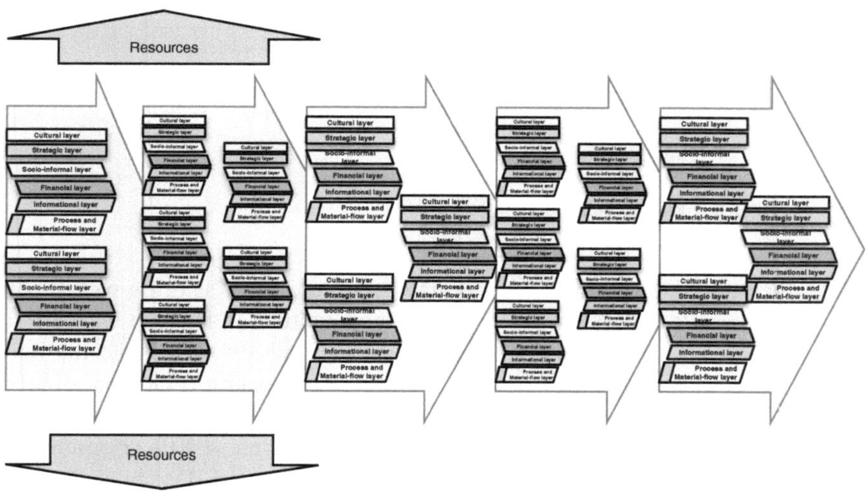

Fig. 4.7 Collaborative process chain configuration by layer described units

versatile value chains. According to the units' respective levels of detail, the units are assumed to organise tasks and activities, Fig. 4.7. Orders, process segments and tasks according to the objectives and the resource availabilities may be executed via team efforts or automated by software agents.

Based on the generic model, including their aspects approaches, i.e. product and resource, the collaborative network manufacturing process may be composed as a configuring collaborative process network, either as a result of human negotiations or as a result of agents' interactions. The introduced aspect layers for network ability do not only allow describing the units and prepare set ups for interrelations; they also allow narrowing down number of properties for smart manufacturing units as expected in Distributed Manufacturing (Fig. 4.8).

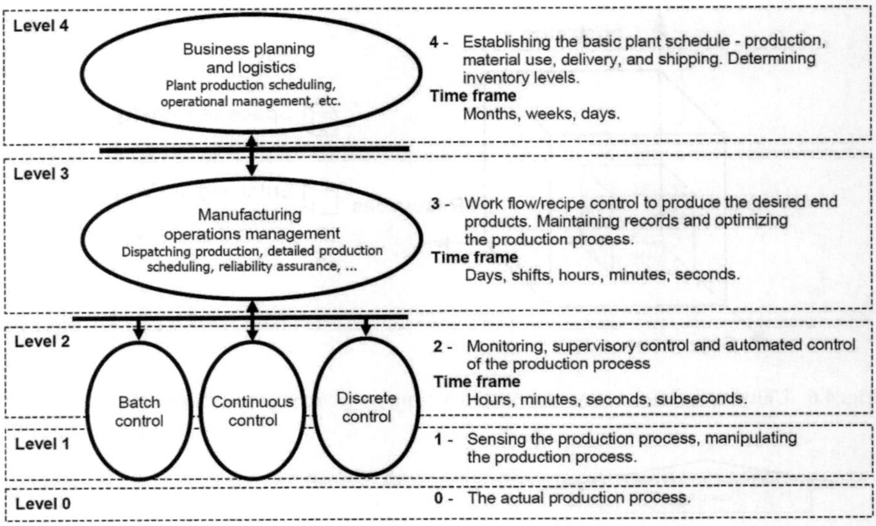

Fig. 4.8 Active level structure according to (DIN EN 62264-1:2014)

4.2 Levels of Detail

Since enterprises are no longer seen as static systems, rather as being part of complex and versatile value networks, logical compatibilities become increasingly important, because the necessity of coalescing the physical world into the information world increases. Information at the points-of-creation (POC) on all levels of detail has to be accessible at the adequate points-of-action (POA) for decision making; (real-time) information may be acquired by (e.g. ERP) programmable logic controllers (PLCs), radio frequency identification (RFID) chips, etc., and adequately processed and presented for decision-making at the POA. As there should be no breaks in flows and semantics while crossing various IT systems, units or levels, standards are favourable, for capturing and feeding in relevant information or acquired data. Thus, e.g. significant subsets of real-time variables have to be provided for control purposes at any network level, so decision making can be performed even with minimum human interaction. Any object has to be able to seamlessly communicate with any other object.

Therefore the definition of a corresponding level logic standard is crucial. It ensures the descriptions of all units and networks in corresponding granularity and levels of detail. As manufacturing has traditionally engaged a number of models addressing different corresponding areas, important models are already able to interact by making use of suitable granularities. In facilities planning and material flow design, level standardisation of objects and flows are common, which apply predefined granularities. In factory planning the respective levels are (VDI 3300):

1. Enterprise network's or site level including all the suppliers' sites.
2. Master plan level, including arrangements of buildings, hangars, storage areas, structural units as production halls, storage areas, and administrative buildings.
3. Layout Planning level with arrangements of departments, machine units, production areas, buffer areas for raw materials, parts and finished products.
4. Work places with single machines, workers places, handling and processing points, as all fold and unfold items, Layout planning may be done in layers of detail.

For corresponding material flow designs, these levels of detail according to (VDI 3300) are standardised as well, along the above facilities planning objects classification:

1. Material flow 1st order is capturing the transports between sites and suppliers as well as customers.
2. Material flow 2nd order is aimed at transportation and handling on the area of the site as well as between different organisational units and buildings.
3. Material flow 3rd order is the movement and handling between departments' shop floors, assembly areas and groups of machines.
4. Material flow 4th order represents process steps within or on a workplace, including handling.

For industrial automation, operations of equipment, processes or systems are displayed for global availability of components and coexistence between wired and wireless technologies.[3] Here as well, a level view incorporates logics of control, which means that the process is identified by a hierarchy of devices and interfaces. The corresponding hierarchical structure for industrial automation is based on ISA-95, also known as the automation pyramid (ISA-95). Hierarchical levels define stages where decisions are taken. The roots are in the production process, and go all the way up to the enterprise management level via intermediate levels. Industrial automation systems comprise various field devices and technologies working in synchronisation; they follow the subordination principle: decisions should be taken closest to the concerned process segment. If the decision range, as given, is not sufficient, decision mechanisms on next higher levels activate.[4] From these mechanisms, 4 levels are defined for all applications. These levels specify closer and ensure communication, dependent on the configuration tasks, choices of detail within a model and between

[3]SLP is an Internet Engineering Task Force (IETF) standard protocol. SLP is a lightweight service announcement and request protocol that allows devices to announce their services to other devices on the network and for devices to query the network for services. In SLP, devices have three roles: service agent (SA), user agent (UA), or directory agent (DA).

[4]Distributed automation is not restricted to wired systems. Wireless communication has been widely emerging in industrial communication due to the increased ease of installation with wireless systems as well as the ability to install systems in locations where wired systems are cumbersome or impossible. Examples of such places are rotating machinery or highly mobile systems. The questions around resolutions of devices, interfaces and logic are harmonised is an interesting one. Thus the scalability of the architecture could be a problem in large-scale smart unit networks. UPnP is a full system configuration and service discovery protocol suite intended for both computers and devices.

models, access to all available data, and the possibility to revisit and to reactivate former results, all in line with the functional hierarchy levels of the recently updated norm (DIN EN 62264:2014).

Supporting systems are complementary, in line with the manufacturing operation management model (DIN EN 62264-1:2014, p. 42)

The levels of detail chosen and the adequate granularity to be applied are resulting from the problem to be solved, for instance on the level of a machine, the level of assembly line or the level of factory, respectively should represent a flow of products on the parts level with machine or a lot size on the level of a line or an order on the level of a factory. These devices are responsible for a variety of functions related to instrumentation, control, supervision and operational management. Here is where distributed manufacturing implementations make a clear difference from smart object use in any other area.

Each level includes specific attributes; information exchange is taking place exclusively with the respective level of the corresponding unit. The top is formed by the extended enterprise level for decisions taken for the entire company. It is followed by the piloting level with manufacturing areas, shops and departments, embedded control and field levels represent the control systems with concrete sensors and actors. Moreover, the level architecture allows conclusions about how many units are installed and which amount of data is involved.

Although the classical pyramid is about to crumble, the logics stay valid, even with the new generations of devices. Based on the new internet technologies, the links between devices, interfaces and logic will gradually dissolve. Putting them all together for distributed manufacturing, there are four main levels of detail as well; for the purposes of DM design and configuration it is hereby defined to distinguish between the following levels:

Enterprise Network or Extended Enterprise level (**EE**): High level reconfiguration, which can be encountered when applying some cases of convertibility and scalability, entails the introduction of a new unit into the MN, the removal of a unit or a combination thereof. A reconfiguration of this nature requires network level supervision in the "ramp-up phase" when the new structure is initiated. In a manufacturing network, which usually involves complex supply chains, the main concern is related to the integration of all members of the supplier and distribution chains, which share a common goal of obtaining market shares through the product realization.

Factory level (**ERP**): (Re-)configuration, which relates to some applications of convertibility, is for example a change in product family that was or was not anticipated when the manufacturing system was designed, but still requires changes and/or setting parameters that can be done by the operators. An operator would provide reconfiguration information to the control system about the new objectives and the physical changes in resource situations.[5] According to standard

[5]ERP-level reconfigurations often can be accommodated in IEC 61131-type PLC programs, but more advanced software functionality (such as IEC 61499 and object orientated programming) could in some cases be beneficial.

VDI 5600, this level corresponds to the enterprise control level, generally supported by ERP.

Shop Floor level (MES): Distributed manufacturing resources have to be dynamically configured for optimizing the problem of reconfiguration and considering alternative process plans for shop floor reconfiguration. The level allows designs of manufacturing execution that is inherently agile. This versatility is achieved through performance parameters: operational agility, and for active structural agility supporting structural changes and reconfiguration. Manufacturing Execution consists of a set of integrated software and hardware components that provide functions for managing production activities from job order launch to finished products. Using current and accurate data, MES initiates, guides, responds to, and reports on production activities as they occur. MES provides production activity information to other engineering and business activities in the enterprise and its supply chain via bidirectional communications. As the actual value creation is performed at the resource level or manufacturing level, the intermediary between the enterprise control level and the manufacturing level is the manufacturing operational level, which is in charge of providing a comprehensive control of manufacturing processes.

Low-level (PLC): Low level reconfiguration has traditionally been accommodated in IEC 61131-type programs in response to commands from a higher level controller. Reconfiguration, which usually relates to customization, is typically a change within process steps that can be accommodated without significant physical changes, as the change in the sequence of operations or an exchange of similar parts. In smart environments, it is provided by the units' interaction-triggered emergence. On this new background, research has been carried out to exploit the SOA paradigm at the manufacturing level too. Prototypical implementations of SOA-enabled devices (e.g., programmable logic controllers) at the manufacturing level have been successfully examined.[6]

It is rather an intensive communication work between the levels and the sub levels that ensures the intended compatibility, which results of self-similarity.

Activities and functions of the units may easily be structured according to the levels of detail (Fig. 4.9), and the decision spaces, if necessary well differentiated according to the relevant network aspects.

[6]In the context of manufacturing enterprises, the implementation of an RTE requires the vertical integration of an enterprise and the realization of closed-loop controls in (near) real-time. Fleisch and Oesterle (2000) propose concepts, which have been addressed by IT within an RTE, suggesting: (i) integration; (ii) automation; and (iii) individualization. In their solution, the integration of an enterprise is realized by means of EAI solutions, capable of connecting various enterprise applications. In general, SOA is part of EAI and seen as a substantial enabler of RTE as it provides flexible and adaptable infrastructures (Zeeb et al. 2007). EDA (Event-driven architecture) can SOA, because services can be activated by triggers fired on incoming events. SOA 2.0 provides a richer, more robust level by leveraging previously unknown causal relationships to form a new event pattern CEP (Complex Event Processing) correlates multiple messages within given time frames.

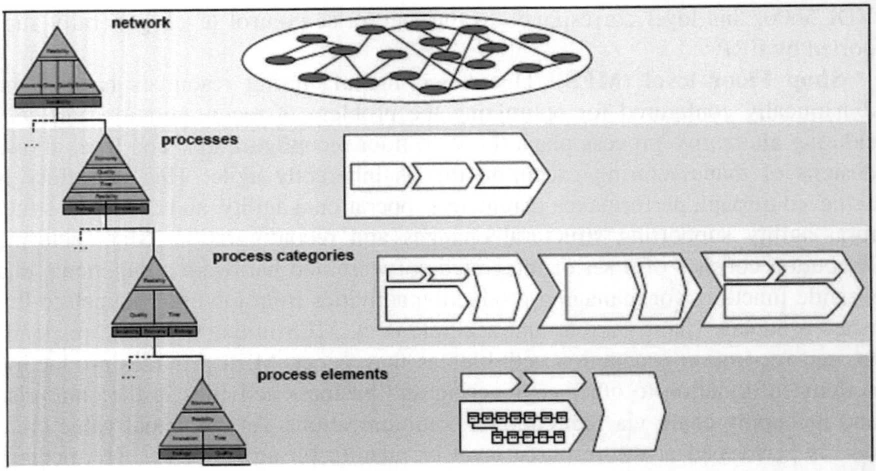

Fig. 4.9 Self similar breakdown of generics for objectives and resulting processes in distributed manufacturing

4.3 Laws and Concurrency Principles

Referring to the generic framework, all core models are subject to principles, laws and rules. These principles must be specific to manufacturing networks. Very general network laws, such as the power law or the small world phenomenon, are valid but not specific enough in this manufacturing context. In the context of collaborative networks and networked manufacturing, more principles had been encountered laying ground for instruments, properties, and tools as needed in cyber units (Fig. 4.10).

In the following, we posit five principles that have been encountered and verified in many examples of distributed automation units and on all levels of detail. These principles command network configuration decisions as well as communication on the field control level equally. We shall call these principles also the network modes or network laws, as they represent key constituents of the theory. The principles support all models, operations, decisions and implementations, as we will see later in the examples' chapters.

The enabling technology essentially consists of powerful devices connected via the Internet and eventually making physical manufacturing and cyber manufacturing worlds coalescent. In manufacturing, where units comprise smart machines, storage systems and production facilities, capable of exchanging online information, triggering actions and controlling each other independently, the guiding phenomenon is the given concurrency of all events, transformations and actions. The resulting subsequent concurrency principles have been verified in intensive research work on collaborative networks, complexity in manufacturing, fractal organisation

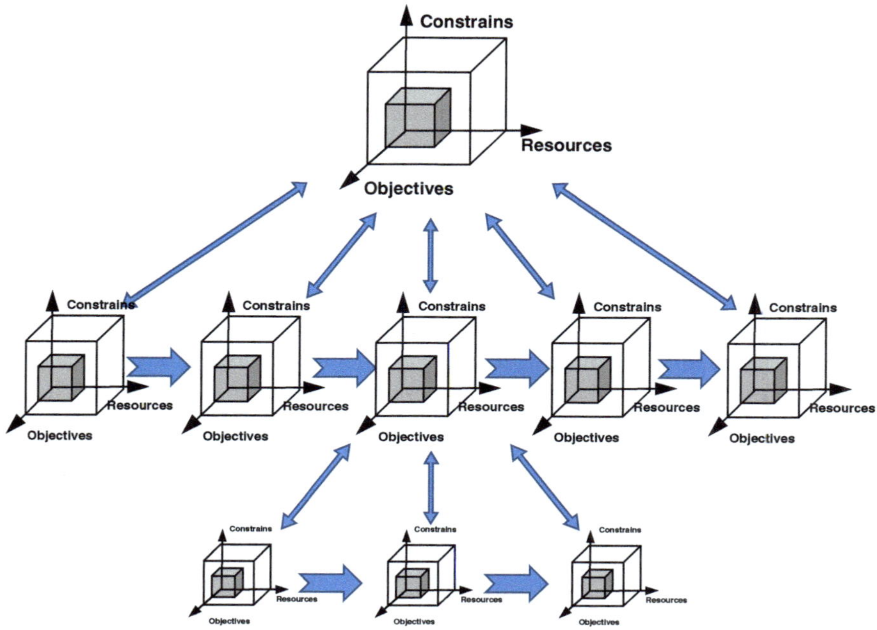

Fig. 4.10 Self-similar breakdown of SoAs onto levels of detail for network level, factory level, unit level

implementations, and criticality thinking; they have been consolidated in many discussions with large engineering communities (Kuehnle 2013).

1. *Behaviour.*

A first principle can be given by specifying the dynamics of the process unit. A unit may execute tasks and actions producing outputs and displaying internal states: a unit exhibits behaviour. The description of the sum of the cited attributes specifies the behaviour of a unit; behaviour comprises all regular actions, which the unit performs according to its technical capability and design. Generally, behaviour also includes the full range of actions made by units, especially in interaction with other units and the environment. In the first place we talk about the behaviour on the physical level of the units engaging powerful logic of cyber physical production systems or smart objects. A unit displays its states and indicators (variables, data) and exposes its behaviour through methods (functions) that act and react to certain events. Process parameters may also express the behaviour of a unit and its interactions with other objects. Monitoring tools enable the users to identify and to interpret events e.g. inter process communication, as long as these events occur at corresponding level of detail of the network units (Kuehnle 2010). The units' behaviour should ensure working according to the objectives, safely, efficiently, and must instigate interacting and communicating in real time.

Irregular behaviour will be detected and either self-healed, or self-corrected. In more severe cases of technical failure or functional disability, alarms may be set and operations self-maintained. More onboard intelligence makes it even possible to better verify, even anticipate a unit's performance in the manufacturing process, after improvement or reconfiguration. Moreover, forecasts on future performances and outputs are possible.

If any unit does not exhibit the adequate mode of behaviour, the required capabilities or is otherwise not supporting the processes in respect to the network adequately, the units' behaviour may become critical, or more precisely: critically behaving units are moving into critical states: they turn into critical a unit that's roles within the network must be checked. The question is to be raised, if further tolerating the autonomy of critical units or accepting units that are (repeatedly) unable to avoid critical behaviour is not a waste of potential and resources. If, e.g. a machine offers loading capacity and capabilities for certain tasks turns out to not being able to perform the task in needed quality or repeatedly proves to be overloaded other network partners i.e. orders are no more inclined to take this machine into consideration.

As models to capture a DM units' behaviour, the Spaces of Activity (SoA) are used. The unit's behaviour, e.g. expressed by corresponding output indicators, may be influenced by the decision space, the result gives input for higher-level decisions on maintaining the unit's self-organization mode or, if the behaviour is highly critical, reducing a unit's autonomy. In cases of severe critical behaviour, network evaluation and order parameters may gain influence on the units' actions ((self) reproduction, (self) destruction, (self) structuring). Should a unit constantly miss important indicators, the network eventually must take control of the non-executed tasks to ensure the total process. Alternatives are provided and e.g. the machine is eliminated from the network.

This biologically inspired concurrency principle copes with most challenges in complex (unpredictable) manufacturing environments linked to self-organization, restructuring and adaptation. It makes manufacturing easier adaptable to changes and eventually commands the global network behaviour by the sum of interactions between units (Bongard 2009). Applied for manufacturing networks, such behaviour thinking provides for the called for "levelled" manufacturing network adaptation procedures.

2. *Parallelism.*

So far, manufacturing has been based on concepts with sharp vertical separation of (systemic) production assets as one time static set up operating in sequential logic. In a network, all this cannot be longer kept up, as the interacting, interlinking units are in continuous configuration mode. This means also that behavioural change as well as physical, logical and interaction issues will work in parallel, in different units and on various levels of detail. If e.g. an order is in process, the smart order itself will keep polling for next process steps and negotiating with units, while another unit is processing the order by changing set ups for customer triggered changes and simultaneously activating other units for getting prepared. This option of doing things in parallel certainly brings new challenges to technology and communication, but, at the same time offers enormous potential for improvement and efficiency. An optimum and well-known way using least resources and time in process optimisation has always been the incorporation of parallelism of all actions and operations. Parallelism automatically brings effects of reducing execution times or increasing throughputs. For lack of machine intelligence and communication, parallelism could not be sufficiently implemented in manufacturing until now, as this principle requires ongoing reasoning about all viable chains of transitions and about ruling it out events that might interfere with others.

Moreover, important mobile wireless devices have become only recently an option. In manufacturing, parallelism is excellently supported by mobile applications; however, their operation requires sophisticated algorithms since it is not sufficient to run just a few services in parallel. Moreover, mobile systems are power constrained, so parallelism profits from strong (wireless) connectivity largely e.g. by shifting consumptive computations to servers or to the cloud. Leading experts state that, generally, more parallel computing systems are underway fully supporting task parallelism and data parallelism, both essential for decentralised and distributed manufacturing applications. For manufacturing applications this allows taking full advantage of task parallelism options on one hand and running all independent operations in parallel on the other. Parallelism adds new dimensions to process planning, for example, by establishing sequences from independent subsequences. Moreover, by parallelism of operations in manufacturing, industrial networks strongly exhibit dynamic forms of communication and coordination that handle non-predictable situations by self-adaptation and self-organization.

3. *Iteration.*

The iteration mode emphasises the fact that there is an inherent and evolving nature to plans and structuring, a comprehensive and recursive optimization process of trial and error. Good network units' decisions will evolve by iteration; the networks gradually improve in economising resources, fulfilling objectives and thereby strengthening the networks in total. Evidently, most promising for planning and decision in Manufacturing Networks seem to be approaches, engaging distributed procedures that proactively iterate and test solutions. Once a configuration has been set up, other options are repeatedly tried out in parallel. For instance, other machines may literally be tried out for load and execution, as the order is processed, or, an order keeps polling and contacting a number of possible next process steps or transports gradually configuring its path through the intended value adding steps. Moreover, iterations result in changes that must propagate through the network structure's stages, instigating continuous process rework. In consequence, network management becomes much more complicated, as the involved units and their behaviours, too, are not stable, but evolve dynamically. Precisely these properties enormously increase manufacturing networks' adaptabilities and strongly amplify differentiations and uniqueness. Developing configuration options and deciding about favourable configurations are therefore inherently iterative processes and not straight-line journeys. Loops back are possible, as identified factories, networks or unit capabilities may not fit; other newly appearing units may give rise to potential new business opportunities, which can happen any time and are immediately detected by real-time monitoring. This means continuous revisions and adaptations of manufacturing networks decisions on structuring, re-linking, or breaking up connections in manufacturing networks. Iterations provide for both, for procedures mapping behaviours into structures (and vice versa) and for manufacturing networks' robustness and stability against uncertainties, operator mistakes, or imperfections in physical and cyber components.

Since the integration of units into processes must be adequately orchestrated for achieving suitable behaviours, it is necessary to evaluate network alignments according to fit degrees or by key alignment indicators (KAI), confirming set ups or triggering configuration altering and iterations. Wherever we look, this principle of iteration increasingly marks the path of setting up optimum process networks in distributed manufacturing; the big game changer here is the upcoming Real-time Enterprise RTE.

4. *Encapsulation.*

Encapsulation, in general, is the inclusion of one thing within another thing, so the included thing is not apparent. In Distributed Manufacturing, this principle concerns the possible encapsulations of abstractions of units (e.g. generic models or task descriptions), transformations (e.g. tasks) and decision tools (Kuehnle 2010). By this, the encapsulation mode enables to configure networks and processes by composing units, either for creating network processes, or for further atomising units to obtain elements vice versa. Self-similarity and compositionality of a unit or a process are directly linked to units or task encapsulations and provide the base for constructing networks from components (Lau and Faris 2007). All encapsulated models of units and connectors may arbitrarily be compressed/broken down resp. fold/unfold.

Models of a unit are accessible through interactions at the interfaces. Any model element, visualised as a unit projection in the framework, may be seen as equipped with connectors (links) for constructing and composing units. All units are loosely coupled with their control including encapsulated connectors, which are used to define and coordinate the control for a set of components (element or composite). The levelled nature of the connectors ensures that correspondent composite units are self-similar to their sub-components; this property provides for the options for network compositions and each unit may encapsulate a number of additional models and methods.[7]

Encapsulated decision processes are crucial for network control. As an important example of the encapsulation principle, a revolving levelled network decision cycle set is subsequently outlined. The decision cycle consists of five interacting

[7]In a composite, by encapsulation, all important properties are preserved in the sub-units. For this reason, encapsulation is strongly propagated as an important feature of newly constructed adaptable components (units are self-similar) and a hot research spot for components' reuse.

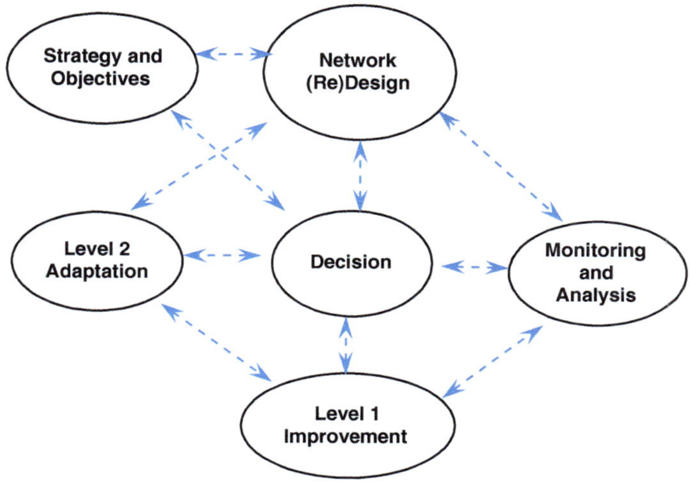

Fig. 4.11 Revolving decision cycle procedure for levelled interventions in distributed manufacturing for gradual continuous network configuration

steps (Fig. 4.11). Starting point is (1) the network strategy and the delineated objectives of the network on all levels inducing network design operations that are (2) monitored and analysed. Critical behaviour will be detected and correction is sought by (3) the improvement step. In the case of nonsufficient improvement, (4) the restructuring case is activated as the strongest reaction. All the steps result in (5) decisions which can either be one of the listed actions or no action. This fundamental decision cycle, an extended ERP decision logic, can be found on all levels of a distributed manufacturing network, varying in parameters, but always in this structure.

More detailed, the interacting steps may be described as follows:

(1) **Strategy and Objectives**

The network gets vision, mission and network draft, which will later be detailed in order to generate the design and the operation parameters. The network strategy has to support the idea that, in order to truly align the structure with business requirements, units float to negotiate and to choose the solutions that best meet their unique needs for harmonisation of networks' and units' objectives is another key task as well.

(2) **Monitoring and Analysis**

This step continuously captures and evaluates the execution of processes and manufacturing operations. It executes by detecting and sensing the current state of the business and operational manufacturing environment, and by monitoring the manufacturing processes for determining, if the manufacturing units' behaviours are acceptable (e.g. concerning performance), and for capturing (unexpected) events and continuously informing on the current situation (e.g. desired, undesired and unexpected events). Monitoring and analysis includes all activities that constantly update the units' potentials, capabilities or availabilities on one hand or that check the network for underperforming units and that notify the network in cases of outages or of alarms (recognised by units' critical behaviour) on the other. Structures, mechanisms and outputs are studied, compared and rated. These analyses may be driven down to sub or sub-sub levels' resource configurations and their contributions to the objectives as well as the SoAs structures (incl. the criticality settings).

(3) **Improvement**

Critical behaviour may be corrected by the unit itself. The unit makes use of its decision space and firstly tries to activate all available resources and to apply to new instruments and tools with the aim of removing insufficiencies. Moreover, optimisation logics may be used to better exploit all available resources e.g. activating additional timeslots, putting more skilled personal or shifting loads along the time axis in order to meet the units' objectives again soon. Especially in cases of breakdowns, self healing, self maintenance and self optimising procedures are applied. These measures should be sufficient to eliminate critical behaviour, temporary criticalities or bridge failures of equipment.

(4) **Network restructuring**

For distributed structures, benefits result from combining complementary capabilities and experiences. Participating units should represent a distributed network of operations. The structure determines, who is involved, how relations, and which tasks and responsibilities are covered by the respective competencies. Participation in the network depends on the complementarities of competencies, scarcity of resources, consistency and suitability of tasks. The restructuring especially concerns the operations round up after removing and or replacing a node as well as selecting additional network participants. The network is (re-)configured to perform and to meet the meet the objectives best. Units and other actors on all LoDs are identified, evaluated and linked into a network structure. Processes are assigned to responsibilities, e.g. by the GERAM procedure. All strategy elements may be broken down to the decisive factors and the respective indicators covering all areas of the networks, which will result in relations of sub objectives or aggregated objectives' systems.

(5) **Decision**

The decision step marks the point where the necessary actions are initiated and executed in order to ensure the network's performance and support its evolution into the intended direction. Decisions may be taken, revised, improved or repeatedly cancelled within this cyclic procedure, i.e. previous program strategy, network configuration, make or buy decision, site decision, decision on process alternative, technology or equipment, etc. are revisited regularly. History and time (as expensive investments from the past) might hinder to execute the necessary structuring decisions immediately. Optimum structures might be achievable, but cannot be instantly implemented for economic reasons or the building up of new competencies will take time. For optimising the network structures it is, therefore, recommended to keep optimum configurations as plans (structure simulator) and benchmark for the actual network. These models should provide for "what if" evaluations and simulated comparisons of indicators that make visible, to what extend the actual configuration space behind desirable configurations and therefore causes "suboptimal" effects on the results.

Any unit component encapsulates all necessary models and procedures. A composite component also encapsulates computation and control (Lau and Faris 2007). For decisions configuring units, special logic, algorithms and methods for integration are given. This concerns the matching of partners as well as the temporary collocation of operations in manufacturing networks. On this base, all units' behaviour as well as all interrelations may be further optimised and planning procedures for the meshed control of configurations (Fig. 4.12), addressing objectives and resources in networked structures, may be established.

5. *Emergence.*

Complexity thinking has means to express varying links and dynamics of interconnectivity, or what in complexity discourse is traditionally termed emergence: arising of unforeseen new structures with unexpected new properties e.g. (Goldstein 1999). Emergence addresses the arising of new patterns, structures and characteristics of networks that are neither really predictable nor fully deductible from antecedent states, events or conditions. For their iterative and gradually evolving character, Distributed Manufacturing configurations and processes, too, are

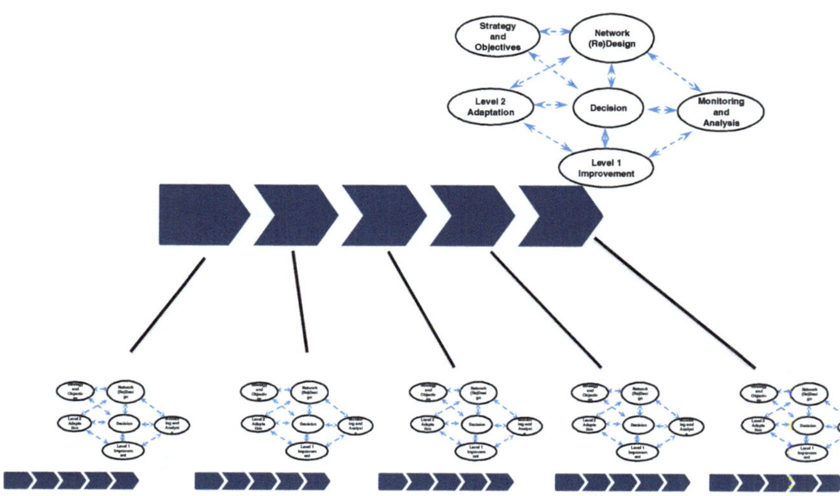

Fig. 4.12 Encapsulated decision cycles with steps and instruments to negotiate and decide on manufacturing unit's output and behaviour on several levels of detail in a distributed manufacturing network

envisioned as emergent. Generally, emerging network set ups are characterised as dynamical, meaning they arise over time, as coherent, meaning they show somehow enduring integration and, occasionally, as ostensive, meaning they appear during a set up evolves. In the smart manufacturing world as outlined, manufacturing processes may therefore be seen as emergent items, precisely corresponding to this emergence definition.

In Fig. 4.13, e.g. alternative manufacturing process chains emerge as a result of iterations and interactions between units. There is no ultimate configuration solution to be found beyond continuous adaptation and restructuring. To state that process chains emerge, however, does not mean that all planning should be abandoned. Rather than deriving outcomes by rigid adherence to preconceived patterns, the key for ensuring good solutions in networks is to focus on creating effective rules for interactions. These rules have two ensure suitable alignments of participants and to increase the likelihood for disposing of favourable network configurations, as aimed at.

Example

In automotive car body shop the model as well as the decision structure are implemented in order to optimize a buffer system and to make the shop more flexible. The objective is to optimise the processes and outcome within a storage area by using adequate commissioning strategies. All objects regularly interact (Iteration) and can simultaneously (Parallelism) be addressed. Every part, every box and also the equipment and the human agents are represented as a Space of Activity (SoA) (Behaviour) and have possible process models available (Encapsulation). The storage area operates on the basis of commissioning parts. Availability parameters as well as performance indicators are monitored and analysed. Criticalities concerning the spaces of activity activate the decision cycle. The revolving logic switches to the improvement step trying out three options 1 Commissioning of parts, 2 Commissioning of boxes containing respective parts, 3 Manual commissioning depending upon their feasibility. For the cases if commissioning of boxes is sufficient, the system operates on this mode, as long the indicators for the parts commissioning are not more favourable. If the commissioning of boxes runs out of

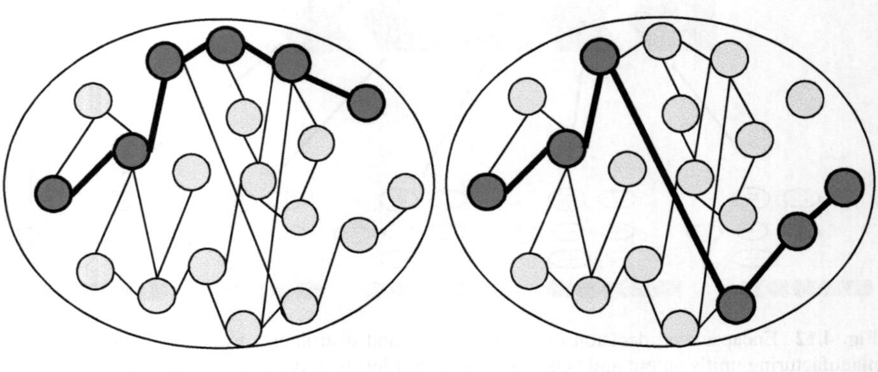

Fig. 4.13 Alternative routings as emergent process chains

feasibility, it switches to manual commissioning. The last resort of restructuring would mean a long term investments decision which is done on a more aggregated level. If all of these options fail, the decisions on the supply of material are given to a higher order decision instance, having in either total process/manufacturing network. The decision cycle firstly assumes the commission to box as to be supplied to the area, as this is not always possible, the option offered different boxes containing their respective parts for picking is chosen, leaving the commissioning to the assembler if this is not possible as improvement, the commissioning is to be manually done by the workers.

The total setup represents a network with meshed controlled elements. The above decision tools of the network appear as encapsulations of these coupled elements and the networked units. Information processing uses different terminology for control levels. Higher levels are in charge, as soon as the regular process, its improvements and additional efforts are unsuccessful to meet the objectives. The cyclic nature of the decision process was not fully implemented at the automotive supplier, as the procedure does not harmonise with the time sliced linear overall planning logics in the network (ERP logic) on the more aggregated system levels. Yet it is an example of a successful hybrid decision set up.

4.4 Information Models in Manufacturing

In DM, products, processes and manufacturing units may reconfigure and interact according to overall objectives for the cyber physical production units, being able to smartly network among each other and also with transportation units. These units may ultimately auto-configure all routings, occasionally supported by humans as long as adequate linking decisions and appropriate granularities are provided for.[8] In this sense, DM means machines talking to machines or units talking to networks of transportation devices; factories and suppliers with all their machines and sensors talking to each other with the aim of smoother processes and higher efficiency. For example, demand fluctuation in Brazil may trigger signals to the devices and parties all the way upstream to a number of globally dispersed suppliers, or a signal of low inventory at a store in Indonesia may ultimately trigger component production at a factory in China that has been identified as the best situated location to deliver.

The smartness of a manufacturing unit (as well as the upgrading of existing technology) is provided, i.e. by virtual machines (VM) as already widely

[8] DM units are part of an orchestration for a wide spectrum of sensor- and source data, models and simulations along with approaches to make information accessible and actionable to the workforce and to the automation systems. With increasing self-aware, self diagnosing and self optimising units, distributed manufacturing is not only expected to reduce downtime and improve efficiencies, but also to self-generate versatile configurations and to execute globally dispersed activities.

demonstrated (Kuehnle and Peschke 2006). These virtual machines appear as the quasi identical twin of the (vendor dependent) controls of units/machines/manufacturing centres on one hand, and ensure full IP access via proper MAC addresses on the other. Virtual machines bring in the full range CPUs with all capacities to store extensive model information far beyond the control functions of the represented units and may be run independently and decentralised. Merely these projections of the original units are used to answer the specific questions about the structure and the behaviour of the manufacturing network in focus. Hence, models are the essential part of the successful application of EDP within any industrial settings with information processing, the engineering and execution of organizational structures, i.e. the engineering and execution of business and business-like manufacturing processes.

We recall: strictly and precisely speaking, we always talk about distinct models of units, whenever we talk about loosely coupled autonomous smart units, spontaneously forming networks and executing processes, concurrent and evolving planning or negotiating decisions. For any planning and design, not the objects themselves but models thereof are used for being manipulated and put into relations. In DM planning, negotiating, interacting and configuring is done iteratively and repeatedly strictly virtually. Only after the activation of certain configurations and actions, physical actions or processes are performed. For this virtualisation purpose, manufacturing network or process configurations and optimisations invoke a number of models and it is natural and favourable to primarily engage the existing, well validated production model world, which has been already established worldwide during the past three decades (Costanzo et al. 2009). Of course, these models need strengthening in two directions:

- valid interrelations between the real units and models and also
- universal interfaces between the virtual projections of these units.

The trends enforce modifications on one hand as well as model extensions, adaptations, and harmonisations on the other. Establishing more comprehensive and more differentiated standards has to be emphasised, as well as forwarding more reference architecture, especially multivendor environments orchestrating service oriented support. Of course, applications of the models, as described above, depend on the availability of decentralised computing power and the respective software. For the near future it is easily foreseeable that even the smallest units will have enormous computing power[9] (Fig. 4.14).

Especially decentralized ICT may carry all the models sketched above, ready to link, to compose, to negotiate and to decide (processed by own or remote

[9]By increasing smartness, abilities to design, to plan and to solve problems are considerably extended, as the respective models and procedures may be run on the control devices. Even the most powerful planning instruments, as virtual and augmented reality, to plan entire factories may be implemented on smaller machine controls, for instance for ramp-up optimisation, as demonstrated in automotive industry already.

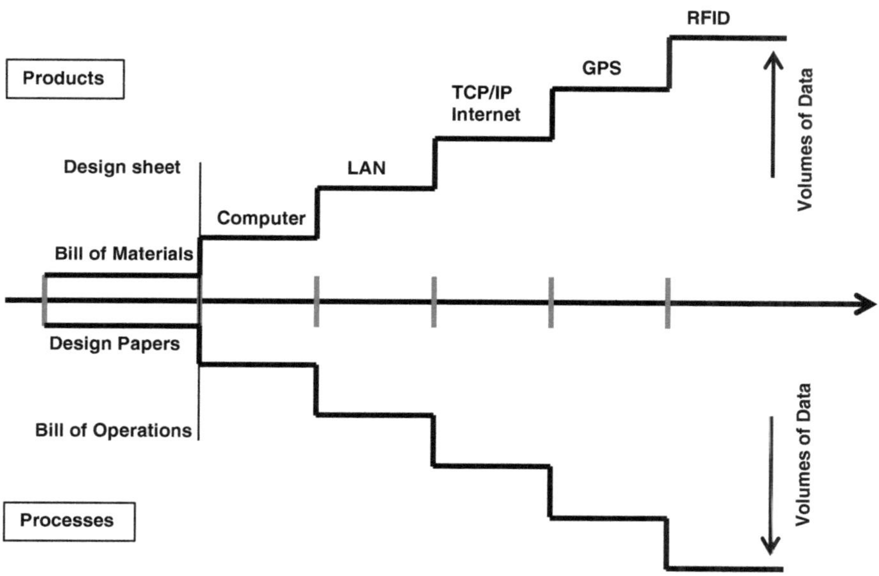

Fig. 4.14 ICT devices and corresponding data volumes and respective model worlds to be implemented, arrayed on a timescale

computing power). Manufacturing then appears as a set up interrelated models; most of the important manufacturing models may be recalled and filled in the framework (Fig. 4.15).

Manufacturing management decisions are basically about adequately matching of product and program dependant requirements concerning resources on one hand, and the available resources under given restrictions on the other hand. In order to highlight this aspect, the models can be structured according to these criteria, the subsequent matching logics and the resulting decision support.

4.4.1 Product Models

Formal mappings of real or planned attributes of products may be called product models, including all specifications of product information as technical documents, artefacts or other representations. Purpose orientation of these models aims at structuring of attributes, specification of geometry and material properties as well as quality and requirements lists. A key information base for manufacturing is, of course, the product design, which, of course includes the bill of materials (BOM) in all different shapes for different purposes. This information base has been used first for programme planning as well as later capacity loads' preparations. With the computer, first CAD systems brought more effective, more

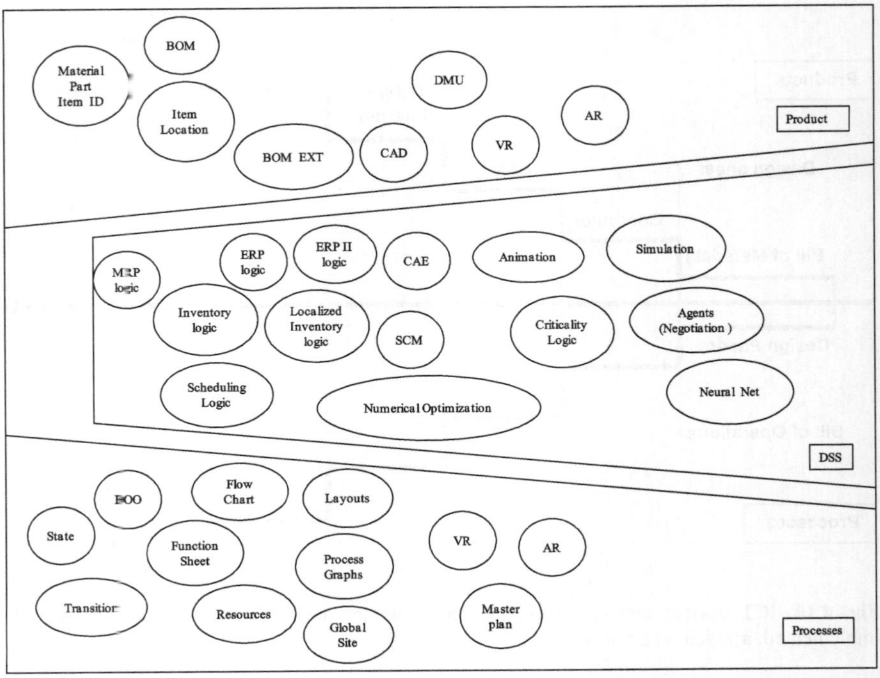

Fig. 4.15 Frequently used manufacturing planning models structured according to their application areas i.e. product, process and decision support system (DSS)

comfortable environments and better updated and maintenance conditions as well as integration possibilities, starting with the CIM breakthrough. In principle, the product information in synthesis with the program plans represents the resource demand side for manufacturing activities. Here, too, information technology sets fast pace for progress and more powerful and better integrated solutions in industries

4.4.2 Process Models

Sequences of operations are called processes and include all activities of analyzing, controlling, implementing and improving, e.g. by harmonised sequences of operations (batch or flow mode) and arrangements of machines and equipment. Planning focuses on sequences of operations that take up all resources, time, space and expertise in order to produce the intended outcome. Any result of elaborate process planning is based on detailed and precise work flow design. For planning purposes two main issues have to be modelled, the task sequences and the operations design.

The task sequences define structures of operations, e.g. assembly transportation tasks, handling activities etc. Operations design specifies the identification of steps (operation), their precise descriptions, machines used, equipment and tools involved, execution times, qualification needed and additional information. For the operations' sequences and time models in manufacturing, the work has always been strongly influenced by graph theory and derived techniques (Coloured graphs, Petri nets) (Falster 1988), the focus has been scheduling of tasks, operations or parts (MRP, Gozinto). These models are mostly applied in sequential steps (capacitated scheduling, queuing models). To support the engineering and execution of operations in manufacturing concurrently, enabling settings to provide the relevant information had been missing so far. With the upcoming devices and the real-time enterprise RTE, online structured information collection, representation, and processing at all levels and for all purposes within industrial settings are available. Decisive new elements for the model world are the tremendous localisation and tracking possibilities of items and objects. These localisation functionalities considerably enhance the preciseness of the resource situations' information, which reduces the necessity of buffers of capacity and inventory, and ensures up to 100 % retrievability of all items, anywhere and anytime. Various sensors allow completely new categories of monitoring improving all databases and providing real-time updates further upgrading the qualities of all engaged models. As project studies (Kuehnle and Peschke 2006) on real-time environments with virtual machines confirmed, well-established manufacturing models are not outdated, did not have to be replaced, they head to just properly be engaged on the base of more precise and real-time data. Therefore, all well established models remain ongoing important representations of smart distributed manufacturing.

4.4.3 Decision Support Models

As already stated, most decisions in manufacturing are about matching detailed demands for planned or available resources or for desired resource situations. Basically, decisions are taken to achieve desired states by means of appropriate transformations, triggered by differences between the actual state and the desired state respectively aimed at. Any decision is considered a rational and deliberate choice of one or several activities and options for action. This means that, dependent on the possibilities of information capturing and processing and the available resources, choices for decision are rational. Any decision process may be described as capturing and processing information by the following stages

- problem description, i.e. especially description of the parameters of the desired states (e.g. via captured data, monitoring)
- determination and evaluation of adequate actions under the given restrictions (e.g. criticality)

- collection of the best choice for action as a solution of the decision situation (e.g. improvement)
- execution of the action as chosen (e.g. restructuring)

Adequate models include information about the states aimed at as well as intensity and priority to achieve them.[10] This includes decision rules and objectives systems forming decision spaces for possible actions to be taken. Restrictions on resources, limiting the actions or producing undesired consequences, are part of the information model as well.

Better information bases, especially wider information bases as provided for by tracking and identification technologies, allow more accurate and more reality near decisions. For the first time, decisions may be taken on the base of plan/reality comparisons instead of plan/revised plan evaluations. We may expect, however, better plans established by simpler, less complex models. On the base of highly improved data for all involved models, the decision support logic may remain basically unchanged, extended by the criticality logic and the negotiation capabilities that are not yet included in the conventional toolbox.

For the total model world to be established, model systems are involved, engaging and linking selected models part of the above categories. Each model should satisfy the encapsulation mode, i.e. exhibit self-affinity or self-similarity features for corresponding levels of details. The corresponding LoDs are outlined in Chap. 3.

Decisions in distributed manufacturing will rather be focused on applying simple models; event and process near monitoring, using online, real time, real network status data will more favourable for good results than further sophisticating models.

References

Alt R, Smit M (2007) Networkability of organizations and business networks. In: Oesterle H, Schelp J, Winter R (eds) 15th European conference on information systems.—St.Gallen, pp 119–130

Bartlett CA, Ghoshal S (2002) Building competitive advantage through people. Sloan Manag Rev 43(2):32–74

Bennet A, Bennet D (2004) The intelligent complex adaptive system, KMCI Press Elsevier, New York

Bongard J (2009) Biologically inspired computing, IEEE computer, April 2009, pp 95–98

[10]First information bases had prepared information for manual decisions, later the computer brought in MRP logics, inventory management programs and shopfloor scheduling. With product and program data, as well as the extended view of value chains, manufacturing company decisions had been supported by large resource information bases as ERP or ERP II. Computer aided engineering and supply chain management software widened up the scope for numerical optimisation, simulation and animation.

Costanzo F, Kanda Y, Kimura T, Kuehnle H, Lisanti B, Singh Srai J, Thoben KD, Wilhelm B, Williams PM (2009) Enterprise organization and operation, In: Springer handbook of mechanical engineering, Springer, Berlin, ISBN 3-540-49131-7, pp 1267–1359

DIN EN 62264-1 (2014) Integration von Unternehmensführungs- und Leitsystemen - Teil 1: Modelle und Terminologie

Falster P (1988) Graph theoretical approaches, computer-aided production management, IFIP State-of-the-Art Reports, pp 97–121

Fleisch E, Oesterle H (2000) A process-oriented approach to business networking. Electron J Organ Virtualness 2(2):1–21

Goldstein J (1999) Emergence as a Construct: History and Issues. Emergence: Complex Organ 1(1):49–72

GRACE (2013) Grace project, URI. http://grace-project.org/

Ivanov D, Kaschel J, Sokolov B et al (2006) A conceptional framework for modeling complex adaptation of collaborative networks. In: Camarinha-Matos L, Afsarmanesh H, Ollus M (eds) Network-centric collaboration and supporting frameworks, IFIP international federation for information processing, vol 224. Springer, Boston, pp 15–22

Kawsar F, Nakajima T (2009) A document centric framework for building distributed smart object systems. IEEE international symposium on object/component/service-oriented real-time distributed computing, Tokyo, 17–20 March 2009, pp 71–79

Kosanke K (2006) Interoperability of enterprise software and applications. In: Konstantas et al (eds), Springer, Berlin

Kuehnle H (1995) L'entreprise fractale. In: Braesch C, Haurat A (ed) La modélization systémique en entreprise, Pôle productique Rhône-Alpes, Paris, pp 263–272

Kuehnle H (2009) Self-similarity and criticality in dispersed manufacturing—a contribution to production networks control. In: Dispersed manufacturing networks, Springer, London, pp 59–76

Kuehnle H (ed) (2010) Distributed manufacturing—paradigms, concepts, solutions and examples. Springer, London

Kuehnle H (2013) Progressing virtualization of production—contributions from distributed manufacturing. In: Disruptive supply network models in future industrial systems, Cambridge international manufacturing symposium 17, Cambridge, pp 19–20

Kuehnle H, Peschke J (2006) Agent technology enhancement by embedded RFID for distributed production control. In: Moving up the value chain, Operations management, innovation and product development, sustainability, service operations, performance management, vol 1. University of Strathclyde, Glasgow, pp 731–739

Kuehnle H, Schmelzer S F (1995) A new paradigm—the fractal view of the factory—the individual in the focus of the factory. In: British academy of management—annual conference 1995, Sheffield, UK Sep 11th–13th, proceedings, pp 278–282

Lau KK, Faris M (2007) Data encapsulation in software components. In: Schmidt et al HW (eds) CBSE 2007, LNCS 4608, 1–16, Springer, Berlin

Lee J, Bagheri B, Kao H-A (2015) A cyber-physical systems architecture for industry 4.0-based manufacturing systems. Manuf Lett 3(2015):18–23

Peschke J, Lueder A, Kuehnle H (2005) The PABADIS'PROMISE architecture—a new approach for flexible manufacturing systems. In: Proceedings of 10th IEEE conference on emerging technologies and factory automation, ETFA 2005, pp 490–496

Piedade RF, Azevedo A, Almeida A (2012) Alignment prediction in collaborative networks. J Manuf Technol Manage 23(8):1038–1056

Rother M, Shook J, Womack J et al (1999) Learning to see: value stream mapping to add value and eliminate MUDA. Lean Enterprise Institute, Cambridge

Sacala IS, Miosecu MA (2014) The development of enterprise systems based on cyber physical systems principles. Romanian Statistical Review nr 4

Zeeb E, Bobek A, Bohn H et al (2007) WS4D: SOA-Toolkits making embedded systems ready for web services. In: Proceedings of the open source software and product lines workshop (OSSPL07)

Chapter 5
Smart Manufacturing Units

5.1 Properties

DM units may be seen as specifications of the IoT, SOs and CPS. Advanced control techniques, cloud computing, emerging network technologies, embedded systems and WSNs are further upgrading devices; the developments may also be seen as an evolution of M2M. The numerous and varying presentations concerning the origin of the devices indicate the rapid convergence of all technologies, so many differences become less and less remarkable. For DM, the developments are anticipated by the introduction of the more specific Cyber Physical Production Systems (CPPS), e.g. since 2013 strongly propagated in the national funding scheme Industry 4.0 in Germany, as these progresses in ICT progressively translate into fast evolving requirements for manufacturing units. Companies will invest in novel technical solutions and to focus their attention on open smart automation platforms for further optimising their manufacturing processes. An essential successful innovation path, which has to be surely classified as disruptive, may be postulated by the smartening up of existing items that are already involved in the manufacturing process. Hence, manufacturing units will increasingly exhibit as equipped with physical and digital objects, upgraded with sensing, processing, actuating and networking capabilities. Abilities, as environment-awareness or self-logging and self-reporting features further augment these items and demand carrying many data about themselves as well as their activity domains.

In order to enable the units to execute the functionalities as assumed, they are expected to exhibit a number of properties, in line with the concurrency principles, harmonised with the novel options and ready to execute all required tasks. A small set of important properties that has proven to be relevant for manufacturing units and all other objects involved in the context of manufacturing, supply and distribution is now detailed. These properties include the features, as found in many requirements' lists for smart manufacturing or smart production, that have

H. Kühnle and G. Bitsch, *Foundations & Principles of Distributed Manufacturing*,
Springer Series in Advanced Manufacturing, DOI 10.1007/978-3-319-18078-6_5

been verified with the first implementations studied by the authors, and will therefore come up again in the examples chapters. The collection is full in line with the technical possibilities, smart machines offer already. Moreover, the properties are the base for further work on the cyber maturity and its technological readiness level TRL of units or companies. With this aim, a maturity matrix within the DM architecture, the Distributed Manufacturing Maturity Model DMMM or D3M, is proposed and applied to the examples in Chap. 6.

Although the discussion is about all aspects, as given by the six layer generics, the properties, subsequently described in the context of DM, of course, are primarily touching the information layer, representing the key aspect for integrating novel ICT capabilities. Especially the cultural and the strategic layer has been covered in the work concerning collaborative networks and extended enterprises already (Camarinha-Matos and Afsamanesh 2008). Like in the extended enterprise context (Schekkerman 2004), maturity conceptions have proven to be extremely helpful in the informational context of distributed manufacturing as well. Synthesising extended enterprise and distributed manufacturing maturity measures appears desirable; however, it would go beyond the scope of this book.

Subsequently, a collection of nine properties, associated with a smart manufacturing unit, is outlined. The properties as outlined are meant on top of other properties and features the units expose already in terms of their tasks and specific transformations e.g. grinding or robotic handling. The descriptions provide more detailed definitions of these properties as well as motivations and preparations of the maturity scales as parts of the D3M that will be proposed in the last section.

5.1.1 Acceptance of Existing Boundaries and Network Participation

Each smart manufacturing unit has to carry its digital presence, uniquely identified in the digital world, which includes ID and network interface address or other application-specific high level naming. Existing boundaries of the DM network must be accepted. This also affects the hierarchies of the (traditional) manufacturing systems in ERP, MES and shopfloor terms with clear responsibilities for factory equipment such as machines or factory sections. Network access and participation is also dependant on a network consensus; high level maturity measures and rules have been provided for in the GERAM methodology. The maturity on the shopfloor level and its technological readiness is measured by the intensity of participation of the network, temporary, frequent or even as a core network constituent. Smart manufacturing units should always retain its original functionalities and appearances; maintenance should extend their physical usages, so it is mandatory to decouple the augmented features from the original unit features. Smart units must support its original functions and properties, even if the augmented cyber part is out of order. Moreover, interactions with smart units should be identical to the interactions with the original object. Mental models cast into emulation may keep the instrumentation implicit (without additional interactions) and will make humans commonly experience that they are dealing with the physical real objects rather than their digital abstract objects.

5.1.2 Modularity

Units are considered modular, if they can be decomposed into sub units that may be interchanged and matched in various configurations. The respective components are able to interact, to connect, and to exchange resources using standardized interfaces. Different from monolithic systems, modular units are loosely coupled. Modularization entails the ability of processes, information systems and products to be packaged as reusable modules that can be (re-)combined with other modules, collectively making up new value-adding compositions.

Modularity relates to the degree of dependency of elements of the module and is realized by allowing loose coupling between modules, implying that modules should have as little interdependencies as possible. In this manner, modular designed objects behave like autonomous network constituents, which can be networked in a relatively straightforward way; it is decisive in which way the units or activities are interconnected. Modularity also implies that, aside local feedback and local decision-making, capabilities are offered for prioritizing task allocation and capabilities which are available for the execution of partial process chains. Standardization is the coordination mechanism of preference allowing modular networked objects to be synthesized in a standard manner, decreasing the need for mutual agreements on interoperability. Naturally associated with modularity is the feature of compositionality, which means that higher level systems' properties can be derived from the local properties of individual components. Compositionality is frequently impacted by strong interdependencies of software and systems adequately designed with embedded higher level properties.

Modularity may be measured by the intensities of couplings, best valued is heterogeneous equipment in loose coupling. As modularity in manufacturing is not a new concept, there are already examples of modules in DM systems, especially in the areas of control systems, equipment design, and human resource development and in enterprise management.[1] Abilities to combine modules, abilities to understand systems of systems and its components, and variably combining these, are keys. Major challenges for modularity are especially the alignments of human resource practices with information systems, so fragmented operations can be adequately

[1]Prominent examples are distributed controls for operations engaging one or more components, equipment in flexible equipment for discrete manufacturing, shop floor autonomy and empowerment of self managed teams, fractals as well as modular interpretations of enterprises in the concepts of virtual factory, virtual enterprise or extended enterprises, primarily aiming at increased agility and flexibility.

supported by human capabilities. The intrinsically heterogeneous nature of modular systems enables to cope with various technologies and tools. In manufacturing, successful use of modularity is mostly based on the ability to align process steps involving different units in order to form viable and efficient value chains by transmitting and exchanging data in a seamless way. It is this kind of modularity that enables to compose manufacturing equipment, which will execute literally every task to be automated.

5.1.3 Heterogeneity

Heterogeneity of units is referred to as the properties of units being composed of diverse elements and thereby using dissimilar constituents. On the way to DM, heterogeneous manufacturing units and their constituents configure networks and have to closely collaborate. Therefore, enabling interactions between sets of heterogeneous ICT devices of different brands and marks is conditio sine qua non in any DM scenario and, therefore, overcoming heterogeneity a central issue in DM, as, due to the variety of involved devices and units. DM is intrinsically heterogeneous. Units or their constituents are to connect and to configure; the respective networks comprise different types of computing units, potentially with vastly differing memory sizes, processing power or basic software architecture. In DM, heterogeneity may therefore be assumed omnipresent, it occurs on all levels and for a number of reasons. On the informational side, heterogeneity may additionally come with different hardware platforms, operating systems or programming languages. On the conceptual level, heterogeneities originate from different understandings and modelling principles for the same real-world phenomena. Basically, two ways of coping with heterogeneous systems can be differentiated:

1. Establishing a comprehensive unified theory
2. Providing abstract data models and semantics.

In this outline, both directions are recognized. Inherent heterogeneity- and integration issues of different components as well as all challenges around are treated with novel unifying network and control theory framework (Chap. 3). The generic layer aspects of the introduced models definitely allow separating heterogeneous connectivity and collaboration issues as well as maintain these in break downs and fold ups. Moreover, heterogeneous networks require permanent revision of network components with emphasis on real-time operations' requirements, so communication and sensing, actuating and processing in meshed control loops are supported by adequate data and object descriptions. Maturities may be differentiated according to heterogeneities of logics and heterogeneities of communication; full maturity is achieved, if both logics and communication may be heterogeneous.

5.1.4 Scalability

The capability to extend or reduce resources in a way that no major changes in structure or application of technology are necessary is generally referred to as scalability.[2] Due to strong worldwide links between the cyber objects and the real manufacturing units, the scalability evidently becomes highly relevant for DM and manufacturing networks. Online connections and real-time data offer so far unseen backgrounds and options for upscaling and downscaling manufacturing assets.

Of course, a main concern is the capacities' scalability, i.e. the facility to increase or decrease necessary resources to efficiently accommodate broadly varying capacity loads. For example, cloud manufacturing gives the cloud clients new options to quickly search, request and fully utilize respective procedures, e.g. for engaging idle or redundant machines and hard tools, also in outside organizations, in order to scale up the manufacturing capacity or scale down own capacity loads respectively and taking in outside loads. Cumulative abilities over capacity, time, space and qualitative scalability indicate increasing maturity. Scalability can be seen as one important feature to realize self organization in DM, as it enables adapting process parameters rapidly in highly dynamic environments. Moreover, in DM, such adaptation processes are gaining importance in plug and work applications. Another field of scalability discussions is the scalability of controls, software and computing power, especially when cloud computing is addressed. Scalability there is most relevant to the commodity background as discussed in remote manufacturing cloud (RMC) setups, e.g. for providing more machines of the same type in different sites or different companies, so large order quantities are fulfilled in shorter times.

5.1.5 Context Awareness

Context awareness, or shorter awareness, is generally defined as the ability to provide services with full characterisation of the current execution environment (Abowd et al. 2002) and any information that can be used to detail the situation of entities (i.e. a person, place or object) is considered relevant for interactions. A smart unit is augmented with various technologies, so, a smart unit is capable of knowing its operational and situational states and of describing itself. Thus, awareness includes both, the units and their interactions.

---

[2]It is measured in dimensions such as administrative scalability, functional scalability and capacities' scalability. Scalability in Manufacturing refers to the ability of a manufacturing system to handle growing or shrinking amounts of loads or usage in a smooth manner by its ability to be enlarged or reduced to fully accommodate the growth or the shrinks.

For differentiation of actions and decision mechanisms for context aware manufacturing equipment, context dimensions may be introduced (Prekop and Burnett 2003):

- External (physical) refers to context that captured by units' interactions or can be measured by hardware sensors, i.e. location, movement, alignment parameters, or strategic input.
- Internal (logical) is unit specific, i.e., goals, tasks, objectives fulfilments, KPIs, improvement effects, operations or processes.

Dependent on captured and monitored data, events or stimuli, a manufacturing object may have to become active. Most important are data used by the encapsulated decision procedures, so the manufacturing objects can adequately respond to monitoring results, e.g. if critical behaviour requires actions. Models to support units with the decision making will take into account possible strategies to guarantee adequate alignment as precondition and for cases in which these strategies could be activated. Performance will be improved through actions and adequate strategies' activation, positively influencing the KPIs defined to measure an objective.

Awareness might be also be supported by secondary infrastructure, e.g. cloud. Aware units offer functionalities for continuously gathering context data and for adapting behaviour. Accordingly, aware systems, as cyber-physical production systems, are intrinsically concurrent, as establishing and running processes are in parallel. Manufacturing information, as specs, work sheets, drawings, or schedule information, is instantly and very precisely available for prompt identification, localisation, processing and communication between actual and planned states and parameters. By definition, the one to one couplings of objects with their cyber model per se exhibit concurrent compositions of computing processes with the physical ones.[3] For distributed manufacturing and the concurrency principles involved, the awareness property is the pivoting point around the decision cycle, as outlined in Chap. 4. The awareness of the context especially allows capturing and verifying a unit's decision space as well as the proper, valid KPI position. Using sensors and actuators for determining current states, once recognised gaps and deviations may be stated, improvements and adaptations may be initiated for exploiting the units' possibilities, expressed e.g. by the space of activity (Fig. 5.1); displayed effects, as critical behaviour may induce actions in the network and in the real world.

To represent the current network states in a model system as well as to bring in modifications (e.g. for optimisation) from the model world into the real world,

---

[3]Accordingly, each of the aware manufacturing objects may carry a number of respective attributes classified into (Abowd et al. 2002), which are continuously verified or updated:

- Identity (unique identifier),
- location (geographic position, proximity etc.),
- Status (or activity) (intrinsic attributes of units, e.g., tool use, processes running etc.)
- Time (local time, timely priorities, ordering steps etc.).

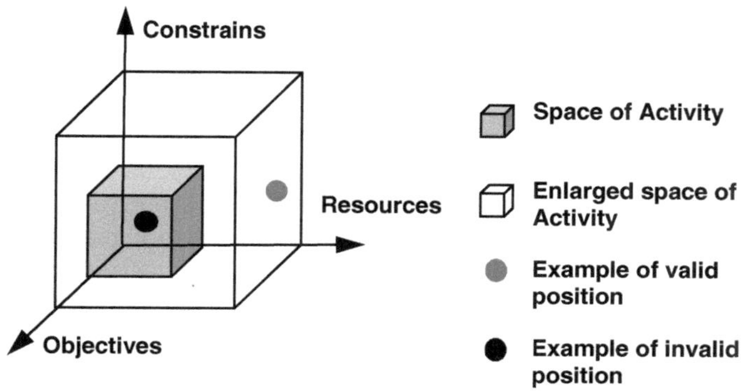

Fig. 5.1 Important awareness contexts for manufacturing units: visualised space of activity or decision space, tentatively enlarged, and objectives' KPI positions

different "network worlds" may be stored as plans and gradually harmonized, so each action in the real manufacturing world may have an effect on the plans and vice versa results in reactions towards the environment. Adequate set ups may be characterised as:

- A set of models that allow us to properly represent the context information at conceptual level. These models are capable to describe information related to possible objective fulfilment, position within the environment, location aspects and behaviour policies, as well as to the users that can interact with the system.
- Strategies and the decision procedures to allow the units to take adequate measures or to anticipate failures and to adapt the plans according to new context data (Serral et al. 2008).

The set ups must as well depict a number of alternatives of possible states and configurations that might be chosen for further optimisation. However, history and time might keep from taking decisions in these directions and, therefore, configurations are to be kept as future options. This notion of configuration thresholds is also called Dual Reality, (Schwartz et al. 2013) (DR, possibly extended to multiple realities MR); the "gradual iterative" decision mechanism behind has been outlined in Chap. 4.

Certain awareness stages of units express degrees of cyber maturity. Moreover, the network environment and the actual situations are also co-defined by other units and the overall manufacturing network as well, especially in cases of critical behaviour. Dependent on reactive or proactive capabilities applicable according to predefined or own awareness patterns are differentiated. For these reasons, the awareness property is closely linked to the autonomy property, which is discussed in the next section.

5.1.6 Autonomy

Units demonstrate autonomy or are respectively called autonomous, if they are able to perform their actions and pursue their objectives without the intervention of other entities (Hasselbring 2000). Autonomy includes the ability to interact or to self-organise in response to external stimuli establishing a positive self-fed loop with the environment. ICT has rapidly contributed to higher intelligence of a number of manufacturing units, so self-control, self-organisation, and eventually full autonomy of factory objects is attainable for most units (Brusaferri el al. 2014). The resulting distributed data volumes and the globally dispersed structures may easily be mastered by company networks or by third parties' services, as the cloud. Autonomous units may now do their communication independently and may decide how to handle interactions with other units and the outside world. The results may be observed as units' own decision making and the formations of autonomous hub organisations with own rules and procedures within a collaborative process or supply network.

Autonomous manufacturing units may have these capabilities to take certain actions as simple as switching from state to state, or as complex as adapting the behaviour by different decision-making or own action plans for self-healing, self organising and self sustaining. Depending of the capabilities, the behaviour and the performance of a unit, the degree of autonomy may vary, which is to be considered as a maturity criterion. The base for the assignment of autonomy to a manufacturing unit is the ability of units to independently define or negotiate own objectives and pursuing strategies to achieve or at least to approach objectives. Moreover, within distributed manufacturing setups, units' autonomies are always limited by the modes, how other network units activate their strategies and how they define their objectives. Alignments of strategies and the harmonization of objectives also include decisions concerning partners' selection, contract agreements, objectives' re-definition and performances as well (not necessarily in the range of a single units' autonomy, rather subject to networks and decisions).[4]

The network units keep own objectives and network objectives aligned with other units objectives in the network, or check modified structures for collaboration by adapting or renegotiating links, restructuring network solutions and confirm or revise objective bundles. Reciprocally, misalignments will result in possible conflicts between the implemented strategies and the defined objectives, jeopardizing the benefits of collaboration or even breaking up processes. Misalignments will therefore be interpreted as critical behaviour of certain units with the risks of overstretching of the resource base. Decisions and actions are taken that certainly reduce or eliminate the respective unit's autonomy, i.e. restructuring or redesign. These influences are directly related to the maturity of a unit or

[4]Such long term decisions and decision preparations are extensively treated in the GERAM framework (GERAM 1999).

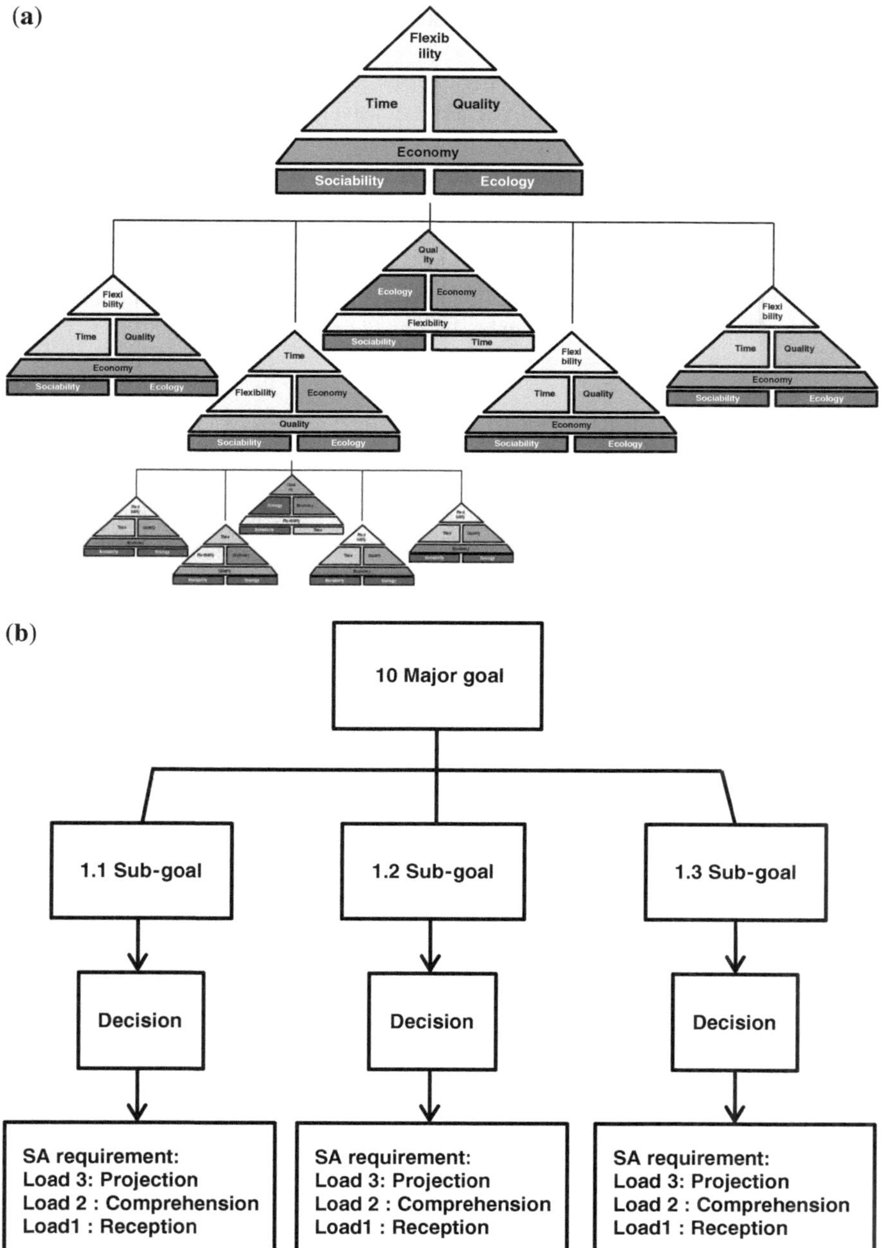

Fig. 5.2 Break-down of network standard objective systems, **a** according to self-similarity principles (Kuehnle 2010), **b** implemented as GTDA software design (Krajewski 2014)

an area; dependent on the power of execution, decision with or without predefined pattern and indicators.

The dependencies between autonomy of units, standard objective bundles, decision spaces and negotiations of objectives are extensively outlined in Kuehnle (2010). A respective commercialised method for assisting in designing and identifying the goals has come up as Goal Directed Task Analysis (GDTA) process, Fig. 5.2a, b. Actionable sub-goals ultimately achieve the original goal. For each sub-goal it must be considered how iterative harmonisation of objective bundles may be programmed in to a units' spectrum to support variable autonomy. Once the business goals and objective bundles of a unit are clearly understood and internalised, favourable manufacturing network configurations can be designed or redesigned.

5.1.7 Interoperability

The property of diverse networks, units or subunits to collaborate (inter-operate) is referred to as interoperability. Interoperability is defined, as soon as operable units are available; operability itself refers to the ability to safely and reliably run a set-up, in line with general and unit specific requirements. IEEE defines interoperability as the ability of two or more units or components to exchange information and to use the information that has been exchanged. Interoperability can be understood as the capability of ICT systems as well as all supporting processes to exchange data as well as to allow sharing of information and knowledge.

In larger contexts, issues of collaboration and co-operation of units appear as communication between people, communication between people and ICTs, and, of course, between different communication technology devices. With a focus on the latter case, several levels of compatibility, and extended interoperability view, are differentiated, according to IEC TC 65/290/DC. This standard puts communication units' features for communication and for applications together in a cumulative scale (Fig. 5.3).

The range extends from incompatible to fully compatible, where the term of interchangeability is used as the immediate level of communication expressing ultimate interoperability. Obviously, the IEC scale indicates interoperability maturity and therefore is taken into account for evaluation suggestions. For facilitating interoperation, the web TCP/IP standard exhibits mechanisms for addressing automatically; the most important implementations are SLP, zero config, universal plug and play (UPnP).[5] Combinations of services and processes, as desired in DM, are e.g. supported by service oriented architecture (SOA). Here, functions are not addressed directly; instead services are requested

[5]The procedure for discovery is another important part, though the most common are universal description discovery and integration UDDI and WS discovery protocol, generally based on XML, Web service descriptions annotated in WS DL and messages encapsulated in the Simple Object Access Protocol SOAP.

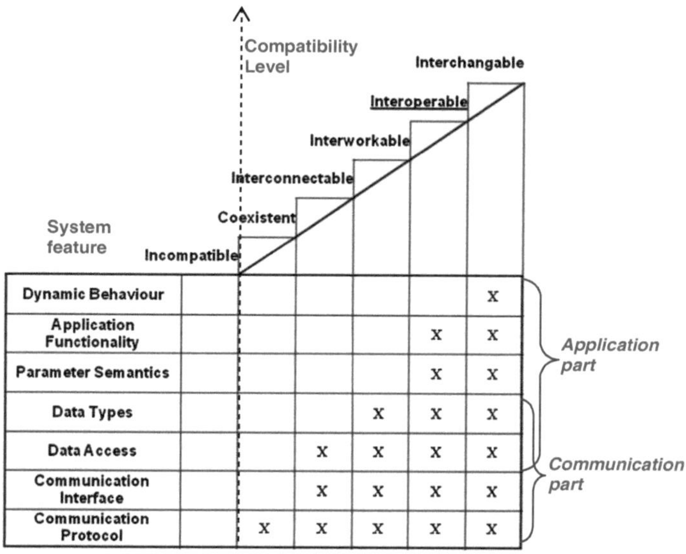

Fig. 5.3 Compatibility levels based on IEC TC 65/290/DC at the same time maturity contributions

via defined interfaces.[6] The service program acts as an intermittent between the client and the provider. SOA is, therefore, an important vehicle for smart communication in distributed manufacturing and a significant step towards new concepts for addressing and hiring services via networks according to usage (pay services), as e.g. offered by cloud providers. A major achievement of SOA is the full incorporation of the concurrency principle of encapsulation for implementing functionalities on generic levels. Encapsulation supports mappings between functionalities of units on various stages of granularity (different levels of detail), by bringing in fold and unfold mechanisms, e.g. hiding or forgetting respectively recalling and exhibiting functionalities in certain situations. SoA set ups and configurations therefore come closest to the Chap. 3 framework structure the outlined theory approach is based upon.

5.1.8 Networkability

Processes in DM may be seen as an inter-related set of functions, ordered by precedence relationships, triggering event and producing observable results (Piedade et al. 2012). Units in DM have to exhibit strongest abilities to network on all

[6]The following features are important: index, representing a collection of services, able for restoration and finding, client for the take up a service and provider, eventually offering service that has been registered.

layers and Levels of detail; networkability[7] may be seen as both, the internal and external ability of units to collaborate, simultaneously considering all process relevant aspects (Oesterle et al. 2000). Networkability is e.g. defined at the DM network level by giving out the rules for alignments of network configuration at all levels of detail of units and subnets. Networkability may be supported by implementing coordination mechanisms that evolve interrelations between units towards networked structures. Networkability of units may be enhanced by sensing and actuating technologies, which capture the global and the local contexts of products, objects and other units, and communication infrastructures, even IT models. In manufacturing, especially process and network decisions are concerned with generating efficient processes. To this aim, smart manufacturing units may even carry factory models, equipment geometries, process and task descriptions as well as interaction and decision models (Kuehnle 2013).[8]

In DM, networkability of units primarily has to promote the configuration of inter-unit collaborative processes on all layers. This includes advanced decision abilities, providing all procedures involved in governing and executing the necessary activities for (re-)designing and setting up new or restructured processes. Networkable set ups result in processes' configurations used as:

- descriptive mapping illustrating performed or running processes for analysing and extracting process parameters,
- prescriptive mapping, supplying anticipated process options for further evaluation and networks evolution, and
- prospective instrument, displaying anticipated eventual configurations for simulation (which configurations should be preferred or avoided).

Reserving horizons of application may be interpreted as maturity signs and—stages that can be clearly differentiated and used as a maturity progress measure.

5.2 Distributed Manufacturing Maturity Model D3M

Of course, the units' properties as outlined are not found everywhere in contemporary manufacturing companies and shop floors yet. For successful expansion strategies there are a number of limitations, questions and management challenges. Core developments in distributed manufacturing that can be extended, however, are implemented in some advanced manufacturing setups already. As it is a completely new field viewing manufacturing as a collection of interacting, autonomously floating, loosely coupled smart units, it would be helpful to have some

[7]Networkability of units may be measured both quantitatively and qualitatively for each of the above aspects. Quantitatively, networkability may be assessed by considering both time and costs, whilst qualitative analysis of networkability addresses the quality of change.

[8]Machine makers have to supply the respective data and tools already, on their website or embedded in the control components.

guidance for evaluating actual and aimed at status under this aspect, to demonstrate options for next steps towards higher integration or to define intended final stages to strive for. Most welcome at this point would be qualitative assessments or statements supported by additional descriptive accounts, possibly accompanied by some quantitative measures. Obviously, the following deliverables would contribute:

- A place to start: It is important to identify a manufacturing networks current state for a unit's evolution, this will help setting necessary actions to achieve the objectives defined.
- A verified benefit of a community's prior experiences by a collection of industry good practices proven by experience to be effective.
- A common language by sharing a consistent dictionary that will assure that every party involved is using.
- A shared vision on the base of a framework for prioritizing actions by providing options for improvement paths, where these are aimed at and what actions could promote progresses.

These are classical items for engaging maturity thinking, as it has been done in other areas. A maturity model is an evolutionary roadmap for implementing the vital practices from domains of implementation and integration (Carnegie Mellon 2006). In order to provide an informed approach for increasing the capability in specific area, these models are used as an evaluative and comparative basis for improvement. Maturity models have subsequently proliferated across a multitude of domains since the concept of measuring maturity was introduced with the Capability Maturity Model (CMM) from the Software Engineering Institute (SEI)—Carnegie Mellon.

Such master plans act as a collaboration rule of networks between units, in distributed networks such as goals, visions, strategies and governance principles; aspects of business operations, organisation structures, tasks, activities and information; aspects of information systems and databases; and the enabling technological infrastructure (Schekkerman 2004).

The maturity model, elicited here in the context of distributed manufacturing engaging smart units, will help to evaluate and improve the collaborative and interoperability practices on the way to full scale distributed manufacturing. From the maturity's point of view, distributed manufacturing is an intricate and interconnected network expressed on the base of the properties of the involved manufacturing units. The ever-expanding universe of ICT devices is technologically pushed, aiming at integrating units and services into coherent large-scale solutions. Units engage in value chains by fully integrating down virtualisations for promoting relationships built around common objectives (similar to the extended enterprise architecture EEA (Davis and Spekman 2004)). Without unlocking units' full properties and effective standards, important distributed manufacturing units may remain stovepipe applications and inefficient processes. Starting from conventional manufacturing and service tasks and process involvement, distributed manufacturing units will interact with

- other cooperating units, continued by
- distributed production across the shop floor or even across sites as well as
- distributed manufacturing controls including several production sites, and
- distributed control applications on several layers and finally with
- newly emerging network controls, characterized by distributed decentralised decision making on all levels.

To properly run such DM architecture, a huge amount of data has to be captured, stored and transferred among the distributed smart units, especially for collaboratively solving a common and higher level control problem. Distributed manufacturing with concurrently led constituents work in real time across heterogeneous environments, across public and private networks, communicating, negotiating and even decision-making. While architectural details may vary from company to company, with the subsequent list of properties as discussed, a number of invariants are stated.[9]

The resulting maturity scale as derived in the discussion properties sections involves technological readiness; it is natural to use technology-based properties as assessment attributes for stating Technology Readiness Levels. The model uses five levels of distributed manufacturing maturity.[10] The total setup of criteria and scales may be presented as the Distributed Manufacturing Maturity Model, D3M. Each model may refer either to a manufacturing unit or to a manufacturing area respectively. The assessment is done by the shopfloor experts and responsible managers in the company (Table 5.1).

Important outcomes are maturity classifications of units and manufacturing areas as well as its maturity progresses. Measures may be observed by viewing the five-point scale in order to assess smart units D3M (technological) readiness. Overall scores may be extracted by summing up the individual units' points scores. In order to give the approach better compactness and more traction, the Maturity Indices may be put into radars; maturities for network may be evaluated by dividing all these sums by the number of criteria (eight) and the number of manufacturing units that are involved in the network.

It may always be questioned, if a suggested grid covers the most relevant aspects and all important points. However, the approach is holistic and the network

---

[9]Whenever collaborative integration is addressed, implementations start locally, gradually and on an exchange base, associating life cycle models, as Generalised Enterprise Reference Architecture and Methodology (GERAM) demonstrating key phases of Extended Enterprise progresses as a whole or the single unit (Kühnle 2003).

[10]There are examples in the literature of models with only three stages (common) and some with as many as nine (Kirkwood et al. 2011). Five was chosen as a suitable number, as it allows a sufficient level of granularity to permit substantial statements on Technology Readiness Levels (TRLs) and differentiation between units maturity whilst still being accessible to the practitioner in terms of making informed choices on units evaluation during assessment.

Table 5.1 Distributed manufacturing maturity model

Unit maturity index	1	2	3	4	5
Acceptance of boundaries and network participation	External	Contacted in negotiation	Temporarily engaged	Full-time network unit	Core network unit
Modularity	None	Homogeneous modules, closed couplings	Homogeneous modules, loose couplings	Heterogeneous modules, closed couplings	Heterogeneous modules, loose couplings
Heterogeneity	Only homogeneous components	Homogeneous communication heterogeneous logic	Homogeneous communication, heterogeneous logic	Heterogeneous communication heterogeneous logic	Heterogeneous communication heterogeneous logic on all levels
Scalability	None	Capacity	Time	Spatial	Qualitative and quantitative
Context awareness	None	Reactive with predefined pattern	Reactive with new pattern	Proactive with predefined pattern	Proactive with new pattern
Autonomy	None	Execution autonomy	Decision autonomy with predefined pattern	Decision autonomy with self-defined pattern	Full design autonomy
Interoperability	Coexistent	Interconnectable	Interworkable	Interoperable	Interchangeable
Networkability	None	Extraction of process parameters	Mapping of running processes	Prescriptive network mapping	Anticipating network configurations' options

view behind allows multiple applications on any unit and subunit level. Putting scores to the single criteria constituting a total observed score may be a useful quick check on the way to smartness and integration in distributed manufacturing.

References

Abowd GD, Ebling MR, Gellersen HW (2002) Context-aware pervasive computing. IEEE Wirel Commun 9(5):8–9

Brusaferri A, Ballarino A, Cavadini FA, Manzocchi D, Mazzolini M (2014) CPS-based hierarchical and self-similar automation architecture for control and verification of reconfigurable manufacturing systems. In: Proceedings of 2014 IEEE emerging technology and factory automation, 978-1-4799-4845-1/14/

Camarinha-Matos LM, Afsarmanesh H (2008) Collaborative networks: reference modeling. Springer, Heidelberg

Carnegie M (2006) CMMI® for development, version 1.2. Carnegie Mellon Software Engineering Institute, Pittsburgh

Davis E, Spekman R (2004) The extended enterprise: gaining competitive advantage through collaborative supply chains. FT Prentice Hall, New York

GERAM (1999) Generalised enterprise reference architecture and methodology, version 1.6.3. IFIP–IFAC Task Force on Architectures for Enterprise Integration

Hasselbring W (2000) Commun ACM 43(6)

Kirkwood DA, Alinaghian LS, Srai JS (2011) A maturity model for the strategic design of sustainable supply networks, ISA working paper

Krajewski J (2014) Situational awareness—the next leap in industrial human machine interface design, white paper, Invensys Systems, Houston, USA

Kühnle H (2003) Enhanced modelling and planning for the extended enterprise. intelligent networking of dynamically interrelated actors. In: Gregory M, Yongjiang S (eds) Exploiting the potential of international supply networks, proceedings of the 8th annual cambridge international manufacturing symposium, University of Cambridge, pp 178–191

Kuehnle H (ed) (2010) Distributed manufacturing—paradigms, concepts, solutions and examples. Springer, London

Kuehnle H (2013) Smart Equipment and Virtual Resources trigger Network Principles in Manufacturing. In: MOIME proceedings, Jakarta

Oesterle H, Fleisch E, Alt R (2000) Business networking, shaping collaboration between enterprises, 2nd edn. Springer, Berlin

Piedade RF, Azevedo A, Almeida A (2012) Alignment prediction in collaborative networks. J Manuf Technol Manage 23(8):1038–1056

Prekop P, Burnett M (2003) Activities, context and ubiquitous computing. Special Issue Ubiquitous Comput Comput Commun 26(11):1168–1176

Schekkerman J (2004) How to survive in the jungle of enterprise architecture frameworks. Trafford Publishing, Victoria

Schwartz T, Kahl G, Applin SA et al (2013) IUI 2013 3rd workshop on location awareness for mixed and dual reality (LAMDa'13), IUI'13 Companion, Santa Monica, CA, USA, ACM 978-1-4503-1966-9/13/03, pp. 115ff, 9–22 March 2013

Serral E, Valders P, Pelechano V (2008) Ubiquitous intelligence and computing. In: 5th international conference, UIC 2008, Oslo, Norway, proceedings 01/2008, 23–25 June

Chapter 6
Networking Tools and Information Models—Selected Manufacturing Examples Applying Concurrency Principles and Novel Units' Properties

The subsequent examples draw from pilot implementations of smart control of rolling mill processes, drop forge and crankshaft manufacturing, and of direct RFID item identification along comprehensive distributed supply, manufacturing and delivery processes. The shift towards event-based or even real time views in synthesis with precise item localisation data reveals so far unseen potential for improving manufacturing management by new ways of decision support. Especially the far advanced rolling mill cases highlight the importance of hybrid decision-making, an interactive men-machine decision finding procedure. This kind of decisions could be important for future smart factory set-ups as well as for conversion phases, where only parts of the equipment display smart properties, i.e. evidently less mature shopfloors in our terminology.

For planning and decision in manufacturing, for operations and facilities planning these game changers even shake up philosophies and shift the priorities in modelling. Whereas conventional planning strongly prioritises processes plans and geometries of objects (machines, buildings, equipment) as decision parameters, now exact positioning and online tracking possibilities highlight real time monitoring and event-based decision-making. Online updates of all data and models, used for planning become possible everywhere and around the clock[1]; the real-time enterprise RTE steadily enters the scene. Installations of adequate systems in manufacturing areas strongly rely on the vast possibilities of capturing data via RFID, NFC or comparable sensors and actuators. Of course all models of products and processes as well as the decision support systems and logics may be installed and provided via cloud. Such novel management philosophy of orchestrating rather than planning is not yet prevailing in companies and may collide with

[1]As conventional planning would include geography and topography either in very early phases of rough planning and conduct strategic conceptions the planning possibilities also in combination with GIS introduce locations, distances and even buildings including story levels at a very early phase.

H. Kühnle and G. Bitsch, *Foundations & Principles of Distributed Manufacturing*, Springer Series in Advanced Manufacturing, DOI 10.1007/978-3-319-18078-6_6

existing routines. The power real-time management however instantly reveals so far unseen improvements and efficiency indicators, so inconveniences should be rapidly overcome by harmonising methods. Implementing respective procedures will certainly induce many installations of RFID data capturing and important investments for smartening up machine units.

The examples are basically chosen from several network levels of detail, as exactly the implementation of breakdowns and fold of network mechanisms and decisions is bound to bring up tremendous potential as soon as the respective fold/unfold logics and procedures are implemented.

Theory setups need verification by solid examples. These examples are provided in the subsequent chapter. The examples use all instruments as introduced in the previous chapters; especially the model threshold for decision-making; the interaction between the aspect layers and iterations are in focus as well as. Steps and procedures are executed in parallel. Moreover, the encapsulated decision cycles, applied across the levels of detail are engaged continuously making routings and status constellations emerge. The major reason for this is that the implemented decision cycle represents a substantial extension of current ERP decision modes, which have not been implemented into these systems yet. All units' behaviour is monitored and all concurrency principles applied. Very concretely, the units' properties and their maturity progresses are evaluated and displayed using the distributed manufacturing maturity model D3M radar, as introduced in Chap. 5.

With these intents, the chapter is organised as follows:

All cases are given an initial display of all applied models and theory elements, using pictogram technique, followed by a short description of the manufacturing case situation. Different patterns of network control around the criticality thinking as the pivoting point are outlined. Subsequently, the maturities of the unit properties are evaluated and application experiences are shortly given. The example chapter is rounded up by an extensive discussion of lessons learnt and of existent and upcoming managerial implications.

All examples validate the approach and clearly verify the development of leading companies into this important direction. All in technical breakthroughs in manufacturing, especially the first steps in Distributed Manufacturing are starting slowly and then grow up to a tipping point, which rapidly unlashes an exponential spread into key branches. The first industries becoming involved are the usual suspects in electronic industry, in metal processing and chemical industries, followed by the big auto makers and automotive suppliers. There are two main settings that profit from distributed manufacturing:

(1) versatile configurations of manufacturing equipment around processes for optimisation of value chains within distributed networks and
(2) sophisticated locally fixed manufacturing units that represent high investments are optimised by virtualisations, and versatility is provided by highly flexible network properties that include several sites.

It is no surprise that this last option for profiting from distributed manufacturing has been exploited quickly by chemical industries as well as by steel- and metal processing industries, as presented in several examples. The upcoming consequences will be enormous, exposing many co-operations within and around the sector with the aim of exploiting all available capacities to a maximum. Cloud technology enables utmost flexibility and versatility of all routings and value networks.

6.1 Distributed Manufacturing Case Studies

6.1.1 Case 1: Smart Reactive Scheduling

In complex manufacturing networks, central decisions become increasingly out dated, because neither current events, nor objective consistencies and other changes do back up the executions of these decisions. Moreover, central decisions for the coordination of dependent production units become less useful with increasing turbulences. The outlined case will demonstrate this problem field by analysing a crankshaft manufacturing area. The shopfloor is organised in four manufacturing units, where one manufacturing unit, the drop forge, is supplying the manufacturing units with parts (Fig. 6.1).

A given optimisation function derives the schedules and the material supply by using classical ERP logic. In cases of changes, be it in demand or be it in objective function, classical ERP functionalities are unable to control and to optimise the processes along the objectives. Thereby, changes are meant to occur during the running manufacturing process and to be various. Under such conditions e.g. change in demand and due dates or changes in lots, varying availabilities of material by suppliers as well as internal events, e.g. machine breakdowns or quality failures, could cause considerable alterations concerning the schedule as well as the output in general. Conflicts arise, which orders should be adapted with respect to due date and size, and the priorities in cases of conflicting objectives.

In this case example, harmonisation is achieved by decentralised coordination of production units. The coordination includes different levels of detail in several manufacturing units.

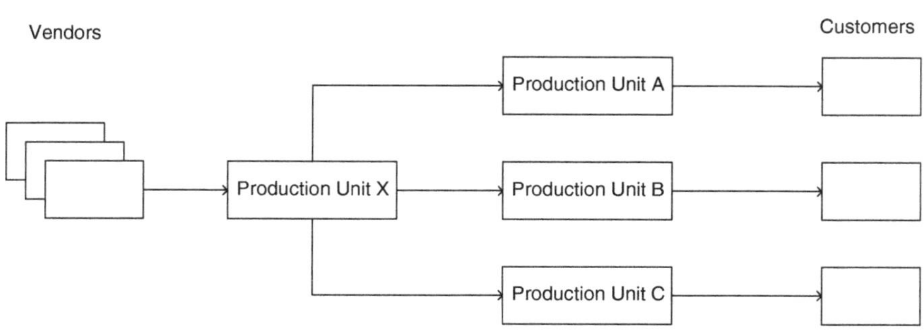

Fig. 6.1 Network structure of the manufacturing area in the case study comprising supply, drop forge and stamp shop units, crankshaft production units, dispatching and distribution

In the logic of the generics, the solution is driven by the informational layer as well as the process layer, Fig. 6.2. On the informational layer, the technical and organisational complications are summarised, so every control unit is organised by local objective indicators, as due dates, inventory levels, and in communication with other network control units. Communication, actively as well as passively are done via central data bases or by using predefined event workflows. The work flows are triggered by top events or newly detected objective indicators that are achievable.

The interactions occur between the process and material flow layer on base of hierarchical, time sliced, levelled manufacturing information systems. The events, the state transition quantities on the process layer are mapped online onto information screens for control. Due to the process nearness, all events may be calculated immediately and continuously evaluated.

The two layer stack has influence on and is impacted by the financial layer; the three stacks represent the total system. Dependent on resource assignments in the network, more or less favourable manufacturing constellations and inventory costs may arise. This first case example stepwise verifies the implemented decision cycle as well as the basis evaluation of the maturity. This case is used to discuss these matters in full-length, the subsequent cases will do this merely in shorter terms.

Configurations in the local controls of the manufacturing units are influenced by maximum inventory levels, resulting from inputs of the subsequent units or from simple physical restrictions. The solution is focussed around a SCM system,

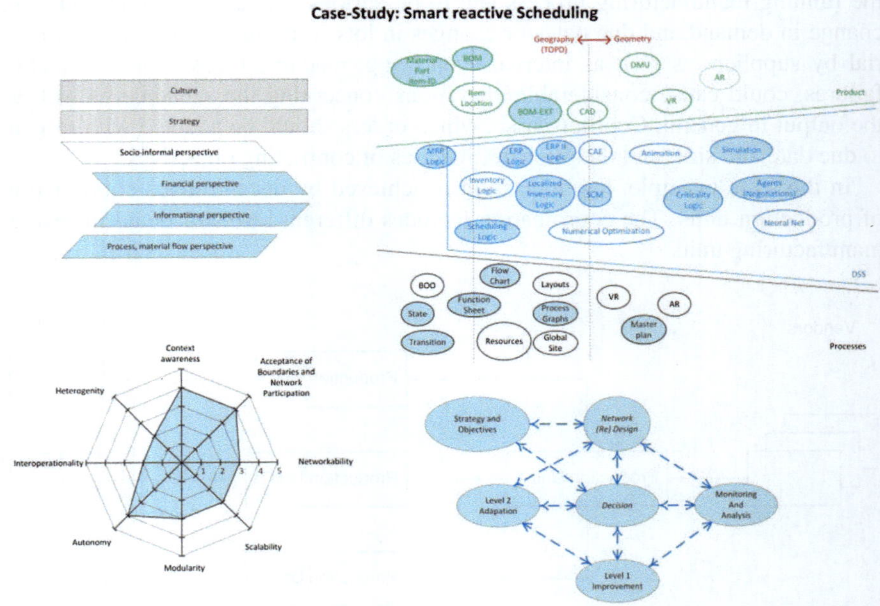

Fig. 6.2 Taxonomy of the smart reactive scheduling case 1 considering involved aspect layers, planning models, levelled decision cycle steps and maturity

a sophisticated scheduling logic as well as simulation. Based upon the demand, calculated from BOM as well as alternative routings, the MRP system details and assigns the manufacturing units. Inputs are the demand, the manufacturing capacity as well as alternative routings in process plans. All elements are provided decentralised for the manufacturing units. In cases of assignment conflicts, solutions are offered by semi-automated procedures according to Fig. 6.3.

In the first step, the predefined solution alternatives are systematically checked, working in flexibility limits. If no predefined alternative does provide an adequate solution, a simulation run is activated starting with the visualisation of the

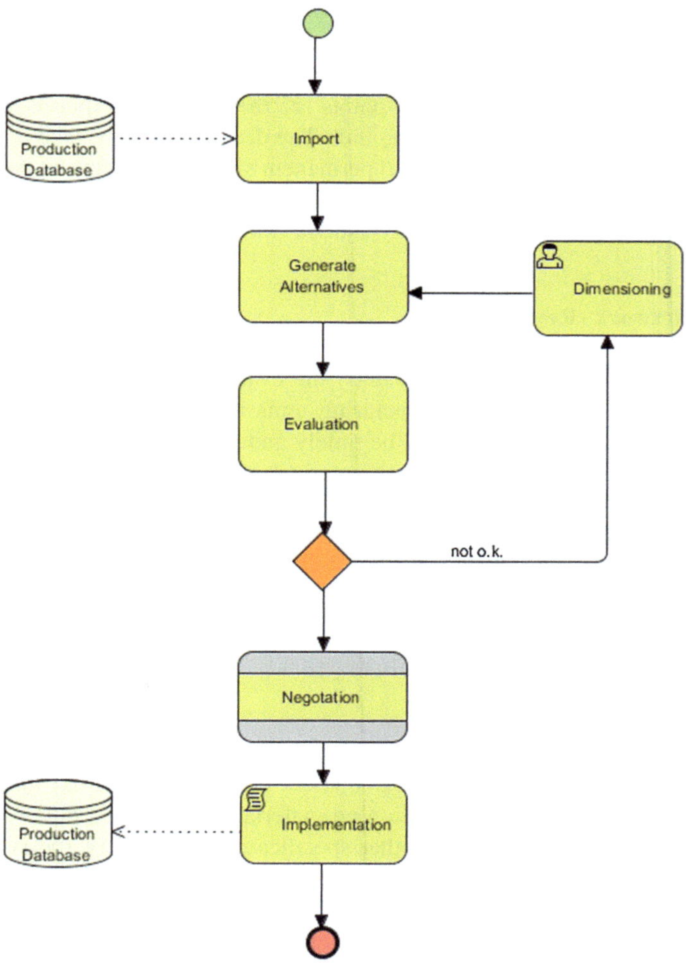

Fig. 6.3 Data flow in the manufacturing area control fed by a mix of algorithms, simulation runs, checks and negotiations around criticality thinking

objectives' conflicts. Next, the flexibility limits, the spaces of activity, are extended in a tentative way and the network re-evaluated under the new conditions. The newly provided solution alternatives in most situations easily resolve conflicting objectives or, at least, minimise objective deviations. These alternatives are negotiated with all involved manufacturing units; the negotiations include feasibility proofs and execution of more solution alternatives' checks, covering both, objective-technical as well as subjective-individual aspects.

6.1.2 Decision-Cycle

6.1.2.1 Strategy and Objectives

The strategy and objectives phase predefines the initial distribution of the loads on the manufacturing units. Reference points are the intended output as well as the capacity of the central manufacturing unit. The decisions will be taken on the base of a master plan and is subordinate to permanent simulation; decision bases are the provided capacities, machines, human resources as well as the forecasted material availabilities, in general via inventory and via storage quantities.

6.1.2.2 Network (Re)Design

Starting from given objective indicators, the evaluation of alternative routings is executed, while alternative manufacturing units are given. Within the spaces of activity, the demands will possibly be timely and lot wise changed or aggregated, according to economic lots or other practical lots sizes. The orders are loaded using predefined parameters and are assigned to the resources on the different levels of detail. The central manufacturing unit, defining key planning parameters, as cargo, oven batches etc., is being monitored carefully and in detail, as its impact on the parameters of other manufacturing units is high and therefore the effects on the total manufacturing efficiency.

6.1.2.3 Monitoring and Analysis

Monitoring and analysis of the parameters is permanent. External indicators are to be included, especially information on material availabilities and order modifications concerning due dates, or other specifications. Internally, the performance indicators of all units, e.g. unit proposed capacity, transmitted as well as fixed process planning times, will be analysed. All changes are evaluated with respect to necessary decisions and alterations. Actions for changes and deviations in its weakest form may be just giving the information, where events simply cause slight modifications in parameters, neither causing completely revised schedules nor reconfigurations. More critical behaviour will cause visible changes, at least rescheduling, but still within the units space of activity, and the improvement mode. The most severe case of permanent critical behaviour induces changes of schedules and reconfigurations, as neither objectives under existing restrictions are respected nor lot sizes and due dates. In this case, the analysis will therefore induce more sophisticated processing.

6.1.2.4 Level 1: Improvement

Changes within the unit provoke level 1 compensations of objective deviations. Dependent on the kind of deviations, e.g. due date, lot size or quality, internal actions are initiated, activating measuring and updating cycles (Burr 2009). For changes in schedule, milestone dates are checked for feasibility. If non-feasibility is caused by outside influences, the super-ordinate units will be informed. Are, however, internal reasons the causes, standard measures will be taken to remove the problems and straighten out the deviations. This is done by using the space of activities' autonomy of the unit itself or in negotiation and interaction with the super-ordinate unit. The responsibility for improvement is fully given to the involved production units, using the granted decision spaces, especially the flexibility, or the resource corridors, as maximum work in process or similar budgets.

6.1.2.5 Level 2: Adaptation

Other than on level 1, improvement, aiming at improving changes in relatively stable conditions, the level 2 changes, compensating major shifts of conditions, are more comprehensive. The adaptation includes reconfigurations of all manufacturing units aiming at the achievement of the given actual objectives, emphasising efficiency and effectiveness criteria. Strong interdependencies of manufacturing processes and process steps bring about that all actions to be taken are subject to sophisticated simulation and the results are visualised, highlighting all simulated key performance indicators. Each adaptation requires extensive knowledge about cross impacts between and within the manufacturing units. This knowledge is focused on the organisational skill of the technical tasks and processes as well as the respective competencies and technical skills.

6.1.2.6 Decision

The decision is repetitively in focus, with choices concerning the next actions. Actions may be defined that precisely sketch all measures for improvement or adaptations, most likely leading to improvements or approximations of objective indicators or configuration measures of the units by iterative adaption. The decision logic is, in this case example, always processed by the performance and control unit that has access to the knowledge of past experiences and results. In this specific case of reactive planning, all demands are harmonised and orchestrated for the different manufacturing units with focus on due dates and output. The resulting indicators are acquitted by the respective units, and the acquitted schedules become approved planning bases for further synchronisation steps. Dependent on the selected automation degree of the planning, acquitting can be done automatically, semi-automatically or manually. In order to avoid irritating backlashes of changes by decisions taken for short range planning, corridors for the spaces of activity are given, especially concerning the resource axis and declared frozen zones.

6.1.3 Distributed Manufacturing Maturity

This case example includes various smart equipment. A number of properties, as described in Chap. 5 are fulfilled, or partly fulfilled (Fig. 6.2). Expressed in terms of the maturity model, the properties' evaluation and observation is given as follows:

6.1.3.1 Networkability—Mapping of Running Processes (3)

Networkability is rated at stage 3, because external and internal processes for all network units are captured and mapped, however the network itself is not modelled on a higher level as a more abstract configuration. All observations are done on the unit level, using internal smartness definitions and all views are implicitly present, but not elaborated or developed as a planning tool.

6.1.3.2 Acceptance of Boundaries and Network Participation—Fulltime Network Unit (4)

All network units accept the boundaries. There is full inclusion into the network; however, no network unit may be excluded from the network without major preparation efforts. All manufacturing units are necessary for the total over all conceivable and offered routings, so all restructuring options are limited or time lagged. For these reasons and arguments, the shopfloor is given stage 4 in the maturity scales.

6.1.3.3 Context Awareness—Proactive with Predefined Pattern (4)

Changes in conditions and environment are realised and captured via predefined observation and evaluation schemes e.g. given milestones or defined escalation levels. Depending on the situation, actions can be taken by choosing out of a predefined set of action patterns, reactively and proactively. Proactively, is especially given on the base of the actions and measures taken already in the context of objective indicators deviations and or missed objectives by means of sophisticated simulation, based on past knowledge, indicators and process parameters. Therefore, the context awareness may be rated at stage 2.

6.1.3.4 Heterogeneity—Homogenous Communication; Heterogeneous Logic (2)

Within the network, fundamentally identical and common mechanisms for data exchange use fixed the data structures, identical location mechanisms and IC technologies. The mappings of the different areas apply various adaptive logics, which is a consequence of the differences in process technology and in shopfloor arrangements for continuous processes. For this reason, heterogeneous logics are just focused on resource allocation strategies and decision logic on the informational layer, so heterogeneity is fulfilled at stage 2.

6.1.3.5 Interoperability—Interconnectable (2)

Interacting network units use uniform predefined protocols and communication interfaces. With respect to the various objects and units, data exchange is done via central data bases offering access to all the components, levels of granularity and access authorisations. Couplings via semantics, as ontology or similar as well as pre-defined data structures are not provided; therefore interoperability is rated interconnectable, grade 2.

6.1.3.6 Autonomy—Decision Autonomy with Self-defined Pattern (4)

All units are autonomous within structurally defined limits. It may be considered decision autonomy for the desired executions. However, this autonomy is not given with respect to the configuration of the units; autonomy is more restricted onto the choice of pre-cogitated actions or action patterns. Units may autonomously define due dates, however the decisions on due dates are not defined at all and who defines due dates is not included in this decision space or autonomy. The given autonomy therefore reaches stage 4.

6.1.3.7 Modularity—Homogeneous Modules, Loose Couplings (3)

Resulting from the limitations in heterogeneity and interoperability, the network units are set up structurally similar. The unit models are coupled loosely, as stepwise extensibility is a crucial factor for successful implementation. All models may be continuously evolved independently from each other. Coupling is principally done via predefined fixed interface parameters using centralised databases. Thus, openness is restricted as not all participants are able to be informed about user authorisations for data structure and the query functions, limiting further evolution e.g. by refactoring. Therefore stage 3 is voted for.

6.1.3.8 Scalability—Capacity + Time (3)

Scalability of the network as well as the network units refers to the time and the quantity of capacities. Fixed localisation of units to a predefined production site and to predefined processes and routines brings about that scaling is not possible or all attributes. Additionally to this systemic limit, deliberate management predefines the focus on decisive process steps in specific company locations and sites. The reasons behind are specific skills and knowledge that are viewed as site dependent, which limits scalability, and therefore is rated at grade 3.

6.1.4 Application Experiences

Within the network, several decision supporting systems are in use, which are now outlined—especially the system for the oven optimiser as well as the shopfloor load and schedulin simulation.

6.1.4.1 Oven Planer

The oven optimiser aims at sketching timing and loading plans for the glowing ovens using their specific process criteria and transformation parameters. Based on the given and the planned orders, all necessary and actually occurring process parameters are evaluated and a batches' plan is established. For the definitions of load alternatives as well as of alternative process parameters e.g. as stop periods, heating periods etc. specific decision tables are successfully used. By means of decision matrix the user may easily adapt and optimise running processes without violating system restrictions.

Loading may be done either manually or automatically. The rule is, to have established loading plans, based on automatically generated proposals that are eventually modified. Existing loading plans may be implemented outside of the frozen zones (Fig. 6.4). Updated feedback and feedback data from the shopfloor reset the status and the time horizon. While keeping planned due dates, batch sizes or lots may trigger local improvements, more severe deviations, i.e. critical behaviour indicates local inabilities, so the criticality escalates to the next decision level.

6.1.4.2 Simulation

Specific simulation is a central element in the decision support system for the short and medium-term dimensions of planning as well as monitoring. With this aim, all relevant systems (ERP, MES) will supply the necessary information and data e.g. manufacturing status, planned material position, planned output,

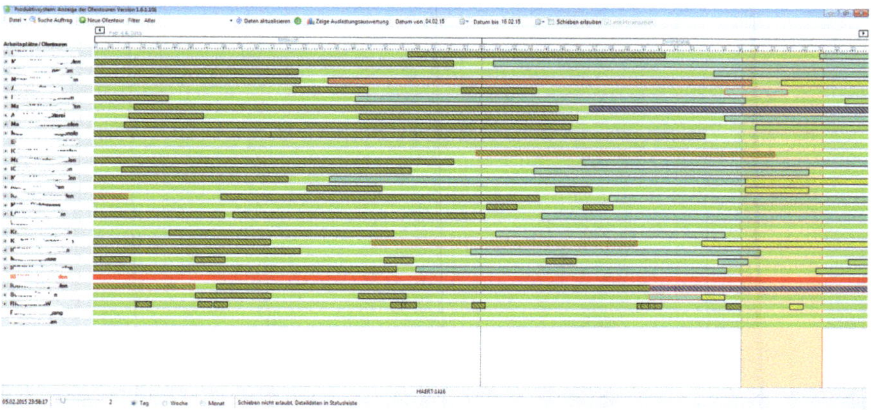

Fig. 6.4 Display—tentative schedule of oven throughput due to widened space of activity by simulation quick check in meshed dynamic order network

capacities. This database, which is locally harmonised and adapted, constitutes the input for the simulation of the planned processes with all assigned resource demands, timely and quantitatively. Scenarios are designed, using varying parameter settings, internally and externally, displaying the involved dimensions (Fig. 6.5). Planning scenarios optimise the system, enormously enhancing the direct objective fulfilments within the flexibility limits, the space of activity, be it by capacity flexibility or by load flexibility.

All evaluations are done on the base of online key performance indicators, before all, it is checked if restrictions are hurt.

Making use of the full decision cycles, which are implemented in each unit, the example demonstrates intensive interaction between the financial layer, the informational layer and the process layer. Monitoring and current evaluation will systematically question the quality of the routings after every significant event. In cases of arising criticalities, the space of activity is tentatively enlarged and the enlargement is tested via simulation. If the simulation result is satisfying, a schedule is implemented with the prospect of gradually shrinking its space of activity, i.e. tightening restrictions or reduced resources' consumptions. If the financial economic figures do not turn out satisfactorily in the simulation runs, the routing is changed i.e. another unit is engaged for executing the process steps. The choice is tested by simulation runs, so all decisions are strictly guided by performance indicators, Fig. 6.6, specific to the plant or manufacturing area. In this sense this way of reactive scheduling may also be called adaptive scheduling, as on online- and corrected data or alert events evoke adaptations in most situations.

This example clearly shows the advantages of concurrently exposing the financial, economic, and informational aspects for displaying process status and evaluations, instigating continuous intensive interaction. Alternative routings' results are

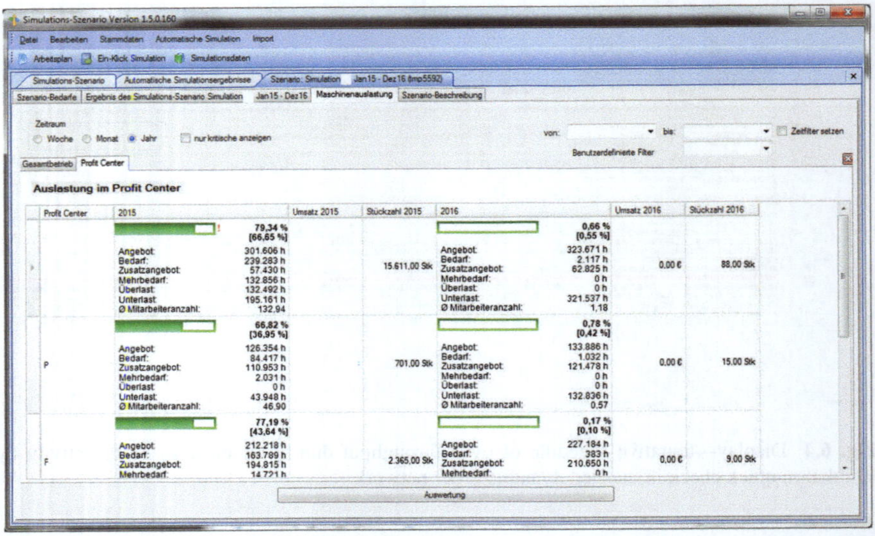

Fig. 6.5 KPI-display of the manufacturing area—feasibility quick check functionality for the objectives' alterations

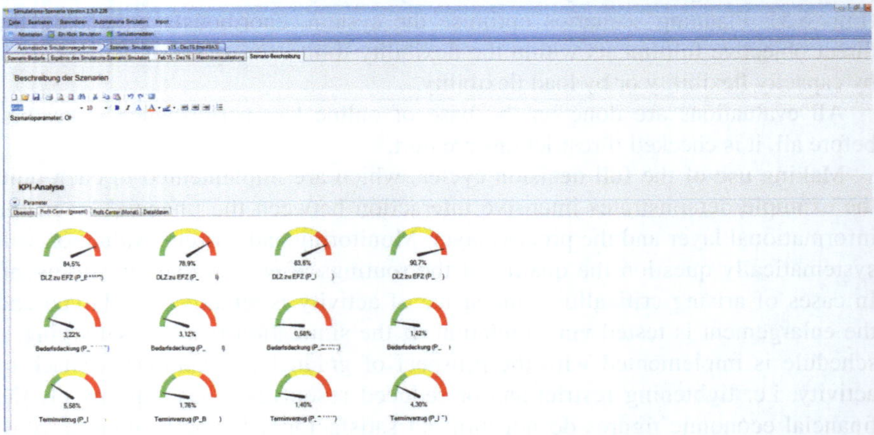

Fig. 6.6 Display—key performance indicator cockpit-chart visualising all impacting changes and tolerances via compatibility quick check functionality

stored, tested for in arising situations and implemented, if the simulation runs are successful. By this means, considerable knowledge is accumulated and kept ready for continuous evaluation and effective decision-making. The high number of engaged models in the taxonomy in Fig. 6.2 is an indicator for this obvious knowledge intensity.

6.1.5 *Case 2: Cloud Computing Supported Tracing of Supply, Manufacturing and Delivery Items*

The subsequent example draws from a pilot implementation for a globally distributed manufacturing network in the chemical sector, comprising several hundred sites, with the aim to enhance the service level for customers and partners by providing highest-quality products via intelligent systems' solutions.

The pilot installation of RFID tagging for improving safety and increasing throughput was done in one of the larger plants with comprehensive supply and distribution networks for a wide variety of products, as plastics organic solvents, products or textiles for paper industries (Fig. 6.7).

The solution has introduced a number of hardware components as handheld devices, fixed readers and antennas, Ethernet connections and PC servers for synchronising software, specifically created from the scratch supporting five main functions:

- reading contents of the tags, linked to the containers
- initialising tags ready for use in production
- saving information about manufacturing status or lots currently in work
- data exporting from tags
- setting up configurations and access procedures.

Applying event based item localisation all along the total supply, manufacturing and distributing process, directly into the cloud; the solution is implemented simply

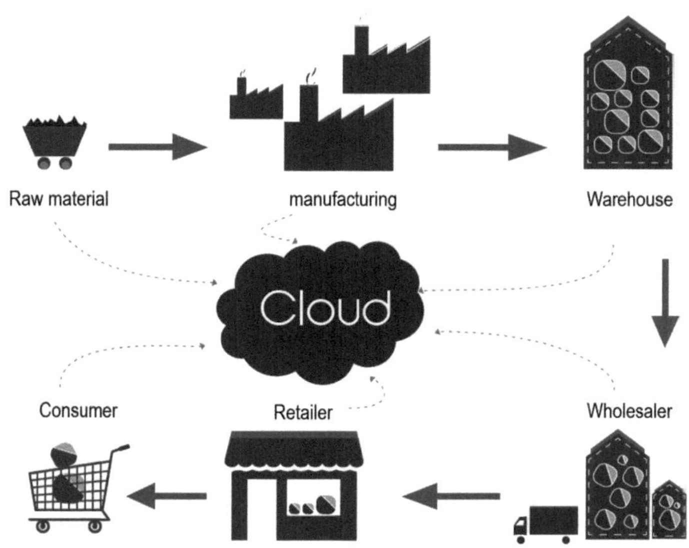

Fig. 6.7 Enterprise wide RFID supported tracking item tracking installation in sketch, small section of all 400 sites on the globe

on top of all existing software support, including installed ERP software. Some ERP interfaces are additionally used to feed in real time data for updating the planning base, because the implemented identification, localisation and tracking inputs have proven to be superior to all existing ERP system data capturing.

6.1.6 Smartness of Units, Granularity and Engaged Models

The installation of such systems in manufacturing initially relies on the vast possibilities of capturing data via RFID and comparable sensors and actuators. Based on powerful positioning systems, the monitoring of all resources, process states of parts and products are provided online around the globe, around the clock. Exact online data for all items involved in the processes may vastly be exploited for improving important dimensions: technical reliability, items' retrievability, capacity management and human resources.

Not only the data but also the models of products and processes for decision support are installed and provided via cloud. It is the decision of the company to what extend redundant information may be stored into the intelligent devices in the process, too. It is another high-level company decision, what data should be given to the public cloud and what capacity of middleware should be exclusively held within the plant or company, for private cloud purposes. The two clouds establish the hybrid cloud. The solution primarily engages procedures for inventory management, especially the statistical inventory control logic SIC with the respective models, equivalent to the widespread ERP logics as shown in Fig. 6.8.

The full decision cycle may be applied to every item. In most cases, the monitoring and analysis is sufficient for real time update and no further action or decision is triggered. The structuring of the network is done according to the total supply and distribution network, as shown in Fig. 6.7. Moreover detailed inventory logic is installed as cloud solution with the usual identification numbering (BOM like), enhanced by status and localisation (geographical data).

Monitoring is executed by the data capturing devices (RFID, fixed and mobile identification). All data are analysed and the inventory updated, including exact locations. Lot orders are released for distinct items according to the ERP exponential smoothing logic.

The **improvement** can be instigated, if the inventory indicators fail or capacity load parameters are underachieved. If indications of frequent failures or frequent asset breakdowns according to indicators are given, **adaptation decisions are taken to modify the structure** e.g. by replacing assets or by changing sites in the structure, depending on the level of detail regarded. As far as the level of detail is addressed, the implementation embraces the network level, the site levels and the shopfloor levels. Specifying localised and logical inventory down to the single item allows matching to any granularity. Such wide adaptability is offered via the power of the globally present cloud solution, able to literally engage all displayed models everywhere and anytime; the model use according to the portfolio in Fig. 6.8 is backed by specific logics, strictly focused on the inventory and item issue (which could of course be instantly enhanced).

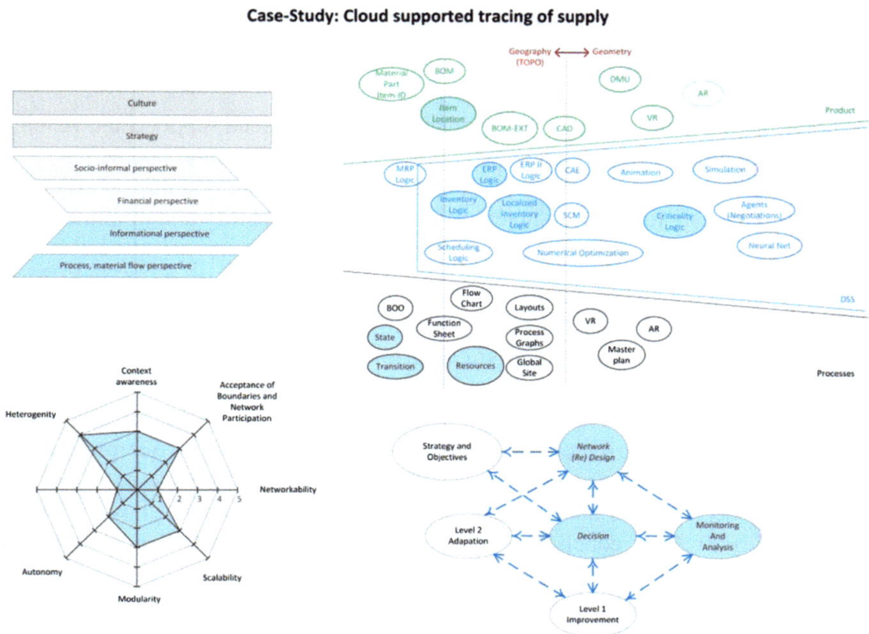

Fig. 6.8 Taxonomy of the cloud supported tracing of supply, case 2 considering involved aspect layers, planning models, levelled decision cycle steps and maturity

6.1.7 Application Experiences

The implementation is running stable, existing and implemented software solutions are kept running unchanged. The additional setups for capturing and processing real-time data, however, enable supply network collaboration across organisational, geographic and systems boundaries with much more informed decision making and shorter response times in fast changing process situations. They also provide collaborative infrastructure to run more centralised and consistent business processes across the organisation's global supply network. The fact that previously installed RFID tags could be reused facilitated the implementation. The combination of RFID and cloud computing helped to keep the investment costs low, so an extremely high return on invest had been possible by

- Improved customer service due to enhanced reliability of product delivery
- Improved operational efficiency
- Increased traceability by access to data via standard IT applications
- On-line technical support.

For example, the trace installation helps an organisation capture and feed event-based information into production planning, order management and distribution systems in order to reduce inaccuracy, mismatches and to improve the speed of decision making. In the case example, the distributed manufacturing implementation accesses and manages information and assets at different points throughout the supply chain ensuring:

- Global tracing to drive operational efficiency with end-to-end supply chain visibility
- Warehouse tracing to optimise space, inventory, equipment and people
- Asset tracing to optimise reusable mobile assets to improve operational efficiency
- Inventory tracing to optimise inventory, maximise availability and minimise costs
- Tracing in retail to increase sales and improve customer service

All responsible managers mainly highlight the enormous efficiency increase and the comfort of the solution. They testify: With our earlier system, managing traceability was a bottleneck in our production line, and even led to delays in delivery of products. Now it is all carried out in real time simultaneously along with the filling of the containers. The system means greater reliability in the delivery of products to our distributors and improves the traceability of our products. A number of other advantages are repeatedly mentioned, as end-to-end visibility of mobile assets, secure data capture and sharing and actionable business intelligence. With 60 % of reduction in operational expenditure and 40 % reduction of loss and underutilised equipment, the key figures display so far unseen improvements. Moreover, with 53 % of increase in order picking and 33 % in overall distribution volume capacity, the customer service has grown up to excellence.

6.1.8 Case 3: Smart Adaptive Distributed Scheduling

Increasing volatility on the customer side as well as on the resource supply side outdate conventional central planning from ERP and time management. Static priority setting in planning, without adaption mechanisms for capturing and evaluating environment influences and without modifications of operations' sequences, will underachieve. In consequence, distributed local solutions of ad hoc nature come up with both, effective and efficient heuristics. Network intelligence, as aimed at, will not be reached by rigidly built-up processes; these set-ups will be just fulfilling short range pragmatic criteria following local muddling through-strategies (Owen et al. 2011).

The case example, subsequently outlined, will propose smart distributed planning and control of manufacturing for a supply network in the metal sheet fabrication, all the way to the distribution. The multistage manufacturing process, each one a separate site, is structured in stages starting from materials, raw coils, from different suppliers depending on the materials. In order to put out final products, a variety of metal foils, manufacturing tasks may be executed on both, in a few cases only on one of the two production units. Most of the manufactured products are processed in a follow up manufacturing unit within a regional company network, respectively another site within the global network. Smaller portions are fabricated and supplied directly to the customers. The finished product will be put out in variants for a plethora of different packaging materials either into buffer stores, or distributed according to customer calls.

Several layers according to (Kuehnle 2014a, b; Costanzo et al. 2009) are touched, process, logistic and material flow as well as informational layer are involved strongly, financial layer and strategic layer are directly interacting, Fig. 6.9. Due to the automation setup, the human interventions are less frequently involved; they have indirect impact by the soft factors as implicitly given by responsibilities or definitions of rules and procedures. The financial impacts are indirect, by company orders and KPI's, as work in process and inventory that are immediately evaluated or predefined. Interactions with strategic issues are included, dependent on the objectives given; the actual demand situation as well as supply status leads to varying objective bundle constellations with rather robust KPI values, not linked to each other, but hierarchically prescribed.

Most intensive real-time interactions between these layers enable real-time leveraging of conflicting objectives. Here, economic financial and organisational aspects are considered as well, however, not in the first place. In conflicting objective situations the adjustment of critical indicators in turbulent environment is not one time static but a continuous overall control procedure. In the general view, all problem positioning occurs impacted by conflicting objectives, notably a balance for economy of scale versus economy of scope should be continuously

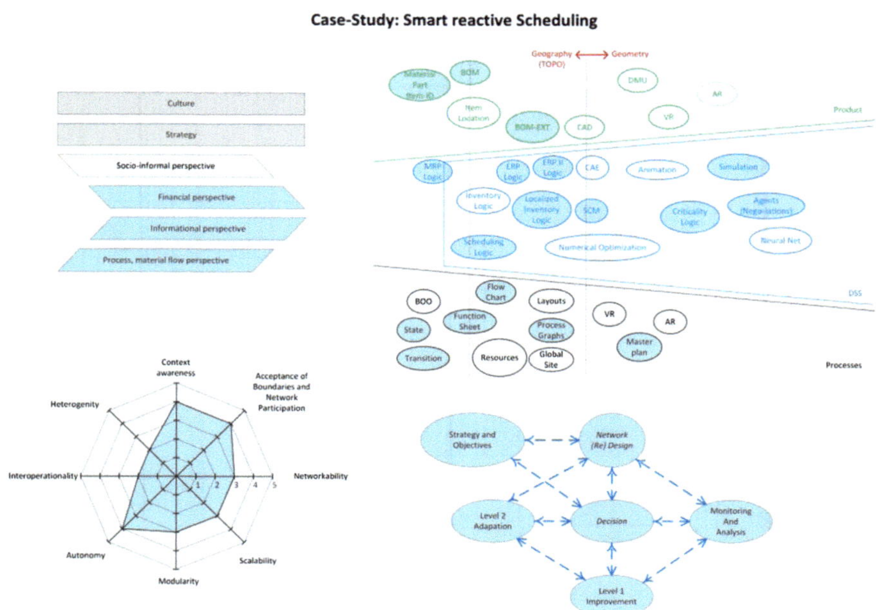

Fig. 6.9 Taxonomy of the discussed smart adaptive distributed scheduling case 3 from sheet metal processing, considering involved aspect layers, planning models, levelled decision cycle steps and maturity

elaborated (Schuh et al. 2009). This means, that the manufacturing units' objectives may be positioned clearly contradicting to objectives of neighbouring manufacturing stages. The first stage may emphasise lot sizing, standard orders and flow orientation; however, the second stage might fully aim at customer orientation by flexibility and keeping due date objectives. The decoupling of both objectives' bundles just by robust buffering i.e. inventory is to be avoided due to negative performance and waste of liquidity. For these reasons, smart and continuous leveraging of this multistage process is crucial for competitiveness, it is a strategic procedure.

On the process levels, the restrictions are consequences of a batch oriented step: Glowing, bringing along extensive process times and forcing to check load oriented efficiency criteria as load percentages of ovens as shown in Fig. 6.10. Contradicting aims are limited storage capacity and limited numbers of production equipment, flexible customer calls and time restrictions in inner factory transportation. Smooth processes are not possible without buffers; however, these buffers should be strictly limited and monitored.

Within the information layer, legacy systems and IT restrictions are prescribed by ERP routines; however, fine tuning and steps for objective alignment are best executed by decentralised procedures. Process logic's requirements of network ability call for communication in peer-to-peer mode (Crowcroft et al. 2004; Yin et al. 2009) that are not necessarily supported on the base of open standards.

From the organisational viewpoint, the manufacturing units are considered to be separate units, hierarchically orchestrated by the next higher network level. Different cultures, found in different plants, have impact on philosophies for the strategies and the respective objective bundles, giving the units deferring

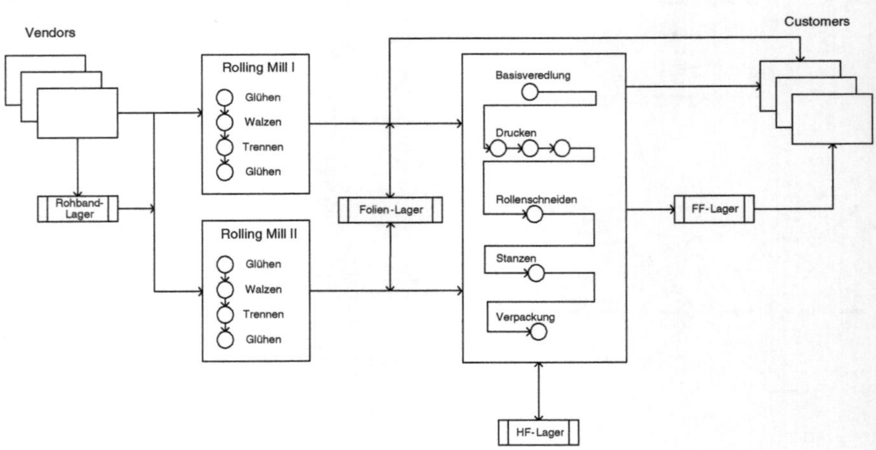

Fig. 6.10 Network structure of the metal processing area of the case study comprising supply, rolling mill processes, finishing, despatching and distribution

orientations. For example one of the plants strictly beliefs in organising the flow by statistical inventory control procedures, whereas other quasi identical plants insist on optimising by MRP. From the network point of view, lead time, efficiency, and delivery on time are only reasonable objectives, as long as the overall network objectives remain robust and sustainable, which is rarely the case.

The solution approach in the case study fully takes up the described threads of the functional hierarchy levels, factory and shopfloor, according to the dissolution levels in the sense of the recently standardised level 3 and level 4 (DIN EN 62264:2014) DIN EN norms. The key functions there are ordering, structuring of orders as well as fine-tuning. All events have direct impact on the higher network levels, so the real-time data capturing and processing within the units is crucial. The levels of detail are not fully modelled, only abstractions of key functional details within the levels are given. The supporting systems are complementary, in line with the manufacturing operation management model (DIN EN 62264-1), where production control and production scheduling as well as product inventory control, material, energy control and order processing are supported; of course, here we fully operate the real-time enterprise (RTE), so the implemented solution exposes important additional information flows, which are not included in the norm.

As always, optimisation of all objectives is impossible within conflicting objective bundles, there is decision space for the adjustment of concrete objective indicators (work in process, excess capacities, due dates). On the base of inventory levels, sketched for all evaluations and KPIs, all variable parameters are continuously checked e.g. lot sizing and lead time. The example highlights interactions between the strategic layer, informational and process layer. Improvements are sought by decoupling of objective bundles and realignment of the objectives on the strategic level. Several levels of detail are involved. The new design of the objectives bundles is supported by numerical optimisation and sophisticated scheduling logic. All network interventions are included as possible actions. The full decision cycle is implemented and considerable maturity is already reached, as the respective radar displays in Fig. 6.9.

The case example clearly demonstrates the superiority of the network view versus the systemic view for these specific value chains. This standard conflict between scale and scope may not be resolved once and for ever by prescribing fixed priorities. Changes in influences and manufacturing status may completely shake up all assumptions so continuous releveraging and reassigning of process tasks within the network is the only way to optimise, fully engaging the levelled decision cycle, as proposed. The maturity radar demonstrates the rather advanced setup of Distributed Manufacturing as described; the results underpin the supremacy of the network view.

6.1.9 Case 4: Hybrid Decision-Making—A Crucial Element in Industry 4.0

The German government initiative Industry 4.0[2] aims at solutions of complex manufacturing network problems by real-time intelligent networking of men, machines objects via ICT (Bauer et al. 2014). This means that more general solutions have to be found and verified by examples. For shopfloor control, generally the descriptions are given based on material and data model synthesis for smart decision support. Key components of such hybrid decisions, as being worked out, are set ups of decision-making, composed of manual semiautomatic and automated decisions and human interventions. Decisions may be taken, based on all these components, dependent on the maturity of the control, the degree of automation as well as the structures' configurations.

Fundamental progresses in manufacturing are proposed on the base of cyber physical systems, respectively their introduction into manufacturing structures. Integration of the cyber smartness may instantly bring about productivity boosts and cost reduction. It is not only the improvement of technical functions by more interrelation, but mainly it is the intelligent linking that will reduce complexity revealing new qualities of problem solutions. Scenarios are seen unwrapping with self-optimising orders passing value chains, and with machines and materials autonomously orchestrating entire process segments (Berger 2014). For implementing such visions and objectives, as propagated, control of complex networks is proposed by use of machine intelligence, respectively intelligent networking. This may happen on the base of autonomous self organising units being supported by software agents, where agents and display proactive and reactive, robust, adaptive, cognitive and social properties. Especially, self-organisation and the involvement of smart objects point into the direction of artificial intelligence (AI), jeopardising the traditional role of humans in the decision processes, e.g. by eventually executing tasks as simple servants.

The example highlights interactions between the socio-informal layer, informational and process layers. Improvements are sought by decoupling of objective bundles and realigning all objectives on the strategic level. Several levels of detail are involved. The redefinition of the objectives bundles is supported by numerical optimisation, sophisticated scheduling logic and simulation. All proposed levelled network interventions are included as possible actions. The decision cycle spares

[2]In German-speaking communities, Industry 4.0 actually is a fashionable buzzword, still lacking of precise definition, and used as a "container term". The term makes allusion to an envisioned 4th Generation of Internet as well as a perceived 4th Industrial Revolution. Backed by postulated advantages in industry competiveness, driven by ITC, interest groups brought up this term for lobbying and for attempts to mobilise government funds. First definitions are proposed with the focus on real-time intelligent horizontally and vertically networked units, machines, objects and ICT systems orchestrated by complex systems dynamic management (Bauernhansl 2014; Bauer et al. 2014).

out the redesign of the network for infeasibilities; criticality thinking is not formalised yet; considerable maturity is already reached, as the respective radar displays in Fig. 6.11.

The manufacturing area is classically structured into production and assembly. The production area supplies all necessary parts for the assembly, where most of the parts and materials as well as purchased parts are put into an automatic store featuring automated transportation to the assembly lines.

Orders may be differentiated in

- customer orders,
- service orders, and
- stock orders.

Stocked-up products are replenished according to ERP forecasts. The manufacturing organisation is classical, structured into general management, area management as well as shopfloor management, with central control and central process planning. ERP, MES as well as automated store and retrieval are applied. The subsequent diagram, Fig. 6.12, sketches the situation.

More detailed, the hybrid decision logic is executed as follows:

Resource monitoring, goal assigned capacities, external objectives as delivery service, flexibility limits, load maximising are considered as input. Long-term customer orders and stock orders are the planning base, the short-term customer

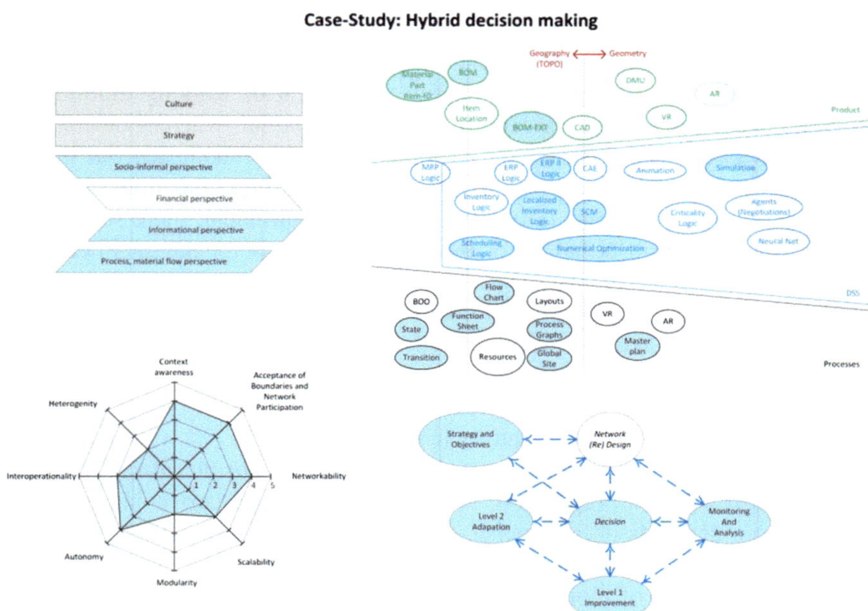

Fig. 6.11 Taxonomy of the Hybrid Decision-Making case considering involved aspect layers, planning models, levelled decision cycle steps and maturity

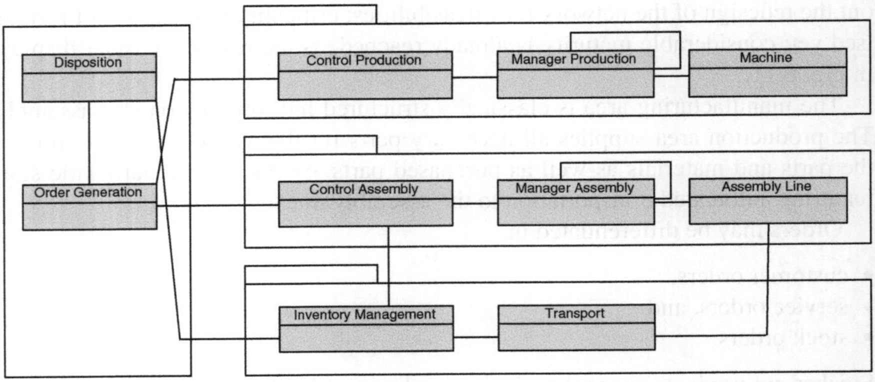

Fig. 6.12 Network control structure of the problem area in the case study comprising automated storage processes, assembly and transportation

orders on top are unpredictable and even more parameters as qualifications for assembly line, limited human resources, vacancies or individual working schedules are in play.

Decisions are made cyclic in stages. The given restrictions will be the input for mathematical algorithms, linear optimising and worker assignment logic, date, individual time schedule, assignment to assembly line. Influences may be played in by simulation i.e. lots, changes in numbers, changes in schedule and so on. All scenarios will be automatically checked for requirements fulfilment; social criteria will be checked leading to asymmetric assignment of working timeslots per worker, enabling to exploit his daily or weekly time buffer, whereas others might be reduced to minimum flexibility. Volatility is mostly smoothened out by controlled assignment of standards orders resulting in more compatible plans. For frequently critical situations, level I Improvement is sufficient; more severe coordination measures will be taken for the assembly unit as for the qualification or restructuring of the assignments, by engaging level II adaptation.

The iteration eventually results in plans for the assembly lines M1–M3 as in Fig. 6.13, including assignments of objective bundles and necessary changes a lot size, or schedule, which will be passed on automatically to the respective units. Proactive planning may be applied, reactive pattern be used, or on demand procedures for unplanned events as volatility in order volumes.

The example proves that unidirectional cost effects are not always adequate for intelligent decision preparation. It is not appropriate to derive capacity loads directly from demands or, vice versa, to define the schedules and loads by existing resource profiles. More dynamic positioning between these extremes definitely supports the finding of more stable and more robust solutions. Moreover, there is still decision space and flexibility for unexpected events. Formalised decision systems are necessary at long-term to bring in more intelligence to the profit of the total network, as purely human taken decisions will not be able to sufficiently

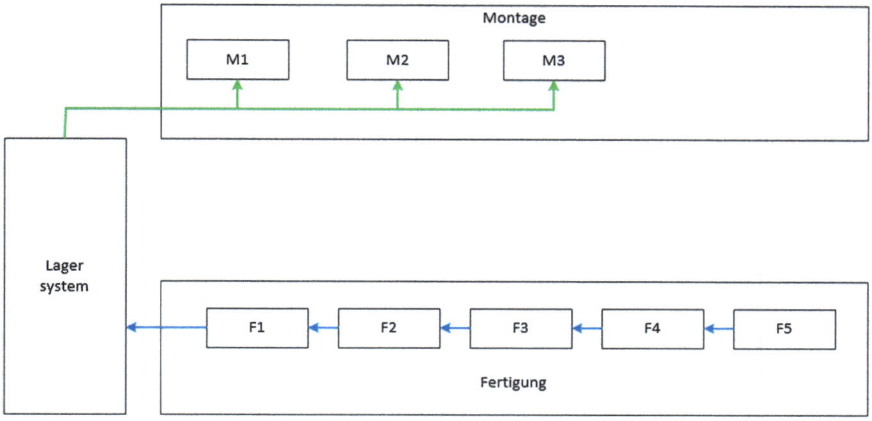

Fig. 6.13 Network structure of the manufacturing area in the case study comprising automated storage, autonomous units in production and assembly

optimise decisions respecting complex restrictions sufficiently. The results are visualised for the responsible person including the fulfilment of objectives. The responsible person now may make use of option for modifying input attributes and for checking by simulation runs. Lot sizes of orders and due dates may be changed, after the respective preconditions, as availabilities of resources have been checked automatically. Social criteria for smooth individual work schedules come in, and adaptations may be done by varying loads of standard orders.

This example of hybrid decision making addresses an important human machine interface and responsibility challenge as prospectively found frequently in distributed manufacturing. Not only control decisions but also Man-Robot interactions will fundamentally change, as increasing machine intelligence will update the strict separation into independent working spaces. Man-Machine collaboration will replace these setups. The key configuration behind the hybrid decision-making problem therefore merits a closer look:

Distributed manufacturing implies that each unit is somehow intelligent or smart, a postulate also propagated in Industry 4.0 (Lukas et al. 2014). Many experts claim that future industry will consist of objects of artificial intelligence (AI), as objects will communicate, exchange information, negotiate and take decisions and therefore control themselves and itself. Raising the question of implementing intelligence is not only be evident in the design phase but also in the operation phase as it is generally agreed that intelligence is not only adequate for setting-up a system, but also for operating it. We give two key arguments for this assertion:

1. Any control is oriented and directed by objective systems that are conflicting. Classical dilemmata in process optimisation may only be overcome by operative intelligence as defined, using real time data and automated decision support. Resulting from the classical dilemma of processes, e.g. lead time versus capacity loads, the problem is growing up to a polylemma in modern manufacturing, as

e.g. value versus plan and scope versus scale have to be included. Such objective bundles may only be mastered by operative intelligence as defined here.
2. Smart units as well as their network configurations expose sufficient intelligence for good decision-making e.g. products with brains (Lukas et al. 2014) making allusions to complex solution capabilities. However such complex solution capabilities are actually not available as automatic decision support, even optimistic protagonists do not predict breakthroughs before 20–30 years time. Some experts claim the need for strong artificial intelligence for complex solutions without human intervention (Kurzweil 2014; Bibel 2011), which cannot be realised within the next decade.

Single units as well as the total network should be able to take sufficiently good decisions now. This still exceeds automated planning as a classical shop floor situation demonstrates: order release without material availability according to plan. Organisation and improvisation will compensate the fact that lacking material is needed towards the end of the task and not in the very beginning; consequently subsets of the order are processed widely differing from the ERP display. Just by citing such examples, no sufficient evidence is given yet for the necessity to integrate human decision skills into the control processes. The main argument refers to the limited rationality of machines (Kahneman 2003; Simon 1991) in the sense of overview rationality, which inevitably links individuals to non-repetitive situations for preventing failures. Moreover, neurobiology findings reveal cognitive performance as being linked to affective components (Ciompi 1997), which calls for the decoupling of human solution behaviour from mechanistic logic and attributes human decisions higher quality and better outcome; reflexivity and intuition prove to be more effective (Reitmeyer 2000).

The desired full intelligence may only be provided by interactions of humans and machines. Intelligence may therefore be considered the constitutional element of this case example's solution; it shall be defined as specific operative intelligence, comprising all capabilities for complex problem solutions. Main attributes of operative intelligence interventions are complex components structure, multiple and conflicting objectives, considerable lack in transparency, high degree of interactions and strong dynamics. Solutions can be designed by use of self organising systems, e.g. implemented as agents (Kuehnle et al. 2011); besides their autonomy, agents have proactive, reactive, robust, adaptive, cognitive and social attributes. Especially self-organisation with the links to the properties of smart objects (Kuehnle 2014a, b; Vasseur and Dunkels 2010) and ICT networks is visibly pointing at artificial intelligence, ambient intelligence, or similar, as the accountable fields. There are discussions, if there are risks that human intervention is only giving inputs or assistance; however in the network context the total configuration of all complements should behave intelligently (Bibel 2011), and not just be intelligently complemented.

For better management of shop floors by intelligent control and networked decision-making, several solution patterns may be listed.

1. the technology oriented approaches towards value networks are to be established, respectively to be altered in a way that automated decisions concerning

structure and improvement may be taken. E.g. Kanban control may be seen as a simple structural decision example (Lödding 2008); moreover automated manufacturing systems of varying complexity may as well be interpreted as automated decision-making systems by procedures trying to eliminate uncertainties, to exploit deviations, to dimension figures or to define alternatives in strategy.
2. Reactive patterns provide more flexibility for removing uncertainties and deviations. Here, flexibility may be differentiated by types into machine flexibility, rewarding flexibility, process flexibility, product flexibility, volume flexibility, expansion flexibility and layout flexibility (Braglia and Petroni 2000). Such wide flexibility interpretation is generally associated to small and medium-sized companies, taking their decisions by human decision-makers and rarely automatically.

The assumption is that hybrid decision making is able to differentiate between manual, semi-automated and automated decisions and generally produces better results. Technical possibilities, as powerful sensors on real-time base, logics in virtual objects, and links to global software solutions providers may instantly provide better solutions by using the subsequent procedure for shop floor control

(a) design shop floor control
(b) check necessity of intelligent control
(c) install hybrid decision model
(d) provide general solution approach
(e) prepare case study demonstrations

The problem solution is to be sketched and resource loads to be defined. Providing for reliable processes is another aim, while depending on organisations, and IT implementations of shopfloor control set ups may vary widely in detail. As a generalised description of the improved shop floor control the subsequent scheme will be applied, where relevant parameters are tasks and their cross-impacts, so the tasks of the control are

- order generation
- order release
- capacity control and
- sequencing.

The objective system includes the four classical indicators: inventory, lead time, load priorities and delivery reliability. Other objectives, as flexibility, cost or similar are in focus. Additional arguments underpin the need of intelligence of units the operation phase.

Hybrid decision making may be implemented best into autonomous units. Autonomous units are defined along the actual decision points, displaying properties of agents, as proactive, reactive, robust, adaptive, cognitive and social. Agents are defined hierarchically, oriented towards re-engineering, deviation detection and reorganisation. Partial self-organisation may optimise, departing from an actual starting point with evolutionary thinking, functional interfaces, goals and input objects pre-defined. For this kind of decisions, differences are to be made between public and private decisions. Public decisions use information for service calls addressed to other systems, private decisions only range within the unit.

Adequate Agents' design may be seen in object-oriented design patterns for units or subunits (Goll and Dausmann 2013)[3] involving:

Encapsulations, (abstraction and information hiding)
Separation of concerns and single responsibility rule and
Interface segregation postulate.

Definition of decision logic, decision **mode** *and decision cycle in autonomous units*:

Departing from service functions various decisions are modelled, it is defined by the logic, how the decision mode of manual, semi-automated, automated as well as the decision cycles (when, proactive or reactive) are built. This leads to more sophisticated requirements for decision implementations, i.e. automated reactive decision must rely on logic dependent on the event time, the time period, real-time, quasi online, or time control. Proactive decisions must enable decision-makers within the units to take robust decisions, which is difficult to model and could rely upon advanced knowledge management approaches, as transactive memory systems' (Lewis et al. 2005) applications for decoding, storage and retrieval (Busch 2008).

Optimisation of decision logic and decision mode

Adaptation and restructuring may be executed internally and externally; autonomous units are internally adapted, if deviations of internal and external objectives trigger improvements. The measures relate to the decision mode, the decision cycle as well as the decision logic, where best procedures for optimisation point at automation and digitalisation, regardless of their actual feasibility (everything that can be automated will be automated, everything that can be digitalised will be digitalised). In line with the distributed manufacturing generics, this optimisation procedure is part of the generic decision cycle as well (Kuehnle 2014a, b).

6.1.10 Application Experiences

This general setup as well as the outlined example demonstrate that hybrid decision logic is easy to implement and extremely favourable in complex

[3]For cooperating autonomous units the relevant design principles have to be extended by (Goll and Dausmann 2013):
Loose coupling
Liskov substitution
Design by contract
Open-closed-principle
Dependency inversion-principle.

environments. All modes, active, proactive and reactive, remain excellently visible and separable. Within the total network, the time cycles are still relatively long for positioning the desired functions. Especially for decision logics with short time cycles, process near, more automated procedures become urgent. Hybrid decision logics as man-machine interactions are in detail difficult to explain, especially in the early setup and ramp up phase. Implementations always require convincing the management, especially for steps for building up respective tacit knowledge. Simple visualisations and decision clearings may be an adequate design.

Moreover, all the new technical possibilities of application specific cloud solution services should be taken into account. The risks concerning data security and confidentiality should not prevent from going into fertile further application fields, as Encryption and big data constituting the most important tasks for providing solutions capabilities within the units. Hybrid decision set ups remain individual, decentralised proactive setups of important resources, will remain difficult to imitate, a fact so far widely ignored, and their transferability will remain widely restricted, which fully ensures lasting competition advantages.

6.2 Lessons Learned and Managerial Implications

What definitely follows for industrial practice is that non-hierarchical views of manufacturing will fully establish. The days of the manufacturing "certainties" of plan, command and control are over; orchestrating the manufacturing networks will replace central, sequential, rhythmic and time sliced procedures by distributed event-driven, iterative, parallel evolving logics. Manufacturing will apply new types of methods and tools, supporting linkage and reconfiguration as well as high level plug and produce, plug and participate, and "hybrid" skills. The examples demonstrate how intelligent networking of partly autonomous units may use harmonising decision support systems, and how high efficiency and best effectiveness may be achieved. Especially the functional encapsulations of flexibility restrictions into units guarantee adaptability for all changes, external or internal. Forecasts by simulation scenarios allow robust action via decision spaces, initiating escalation strategies, dependent on the criticalities in cases of insufficient behaviour. Decentralising brings complexity reduction for the manufacturing unit within the network, opening up spaces of activity and decisions e.g. for dynamic loading between units. The tracking case 2 does not even make use of the full decision cycle. Capturing, monitoring and analysing are the main steps together with improvements. Huge potential is realised by decisions simply on the base of excellent data and prospective statistical inventory control logic (SIC) in the sense of replenishment.

The total network potential, however, is still constrained with respect to reconfiguration possibilities. The reasons are especially identified in the predefined process routines and steps and on the other hand in the lack of readiness to substitute manufacturing units completely or partly. In cases of capacity expansion of the value chain and increasing integration of new units, reconfiguration ability merits enormous attention. Shortening measuring cycles reduce time lags and reaction times enhancing the reconfiguration and the reaction speed of the networks in total. Increasingly smart and RFID or NFC labelled objects will further speed up this development, so easy reconfiguration options will be the rule. Another technology path is certainly the progress in automation of decisions offering options for cutting edge competencies and high tech resources. For planning and decision in manufacturing, for operations and facilities planning these new options simplify and shift priorities in modelling. Whereas traditional planning focused on processes plans and geometries of objects, as machines, buildings, equipment, now, precise localisation and tracking possibilities highlight value chain monitoring: real-time update of all existing data models used for planning, control and execution is given, around the globe, around the clock.[4] Of course all models of products and processes as well as this as decision support systems and logics may be installed and provided via a cloud.

All cases still make use of traditional ERP systems, outcome of another philosophy; the two planning philosophies are not harmonised and synthesised yet. Distributed manufacturing with new technology and novel procedures, however, provides for easy installations of novel data capturing; everything can be run in parallel, as the necessary computer powers may be fully virtualized for the companies and supplied by cloud providers. The quantum leap is real time online monitoring, including precise localisations of all items, which gradually reconciles the contradicting planning views. The global player from the chemical industry from the case example e.g. of course makes use of these real-time data via standard ERP interface for the enterprise wide system and just applies the conventional logics on the base of accurate data; more potential will certainly be unearthed by introducing concurrency, criticality thinking and continuous decision cycles on all levels.

Most improvements are achieved by close and intensive interactions between the informational layer and the process layer. Within the process layer, just a few resources are monitored yet. The full effect may be expected when all involved resources, e.g. machines, transportation systems, human resources, are included; progresses on this path may be followed by applying the maturity radar. The cloud may further multiply the effects by being globally available and offering enormous (remote) computing power.

It could also be shown that hybrid decision logic in complex environment is extremely useful. In the selected case example of scheduling active, proactive and reactive patterns go very well together. However, here too, the function is positioned at a decision point where the reaction times are long; for decision with faster cycles, higher automation is required. At this stage, Man-Machine Interactions with hybrid decision logics are difficult to explain. Convincing

[4]As conventional planning would include geography and topography either in very early phases of rough planning and conduct strategic conceptions the planning possibilities also in combination with GIS introduce locations, distances and even buildings including story levels at a very early phase.

arguments have to be found for the management as well as for the teams and automation takes time to be built up. Visualisations and quick verifications of the decisions are considered an important research path for the future e.g. Ecological Interface Design (Kilgore and Voshell 2014). A good possibility is to make use of the ever-growing technical options are again the cloud solution engines. However, data security issues and privacy requirements are hindering aspects, but the data may be abstracted and neutralised. The most important finding, however, is given by the novel human structural solution competency within the units; proactive set-up of these competencies does not only provide for better results but also for long lasting imitation protection and very limited transferability, ensuring outstanding uniqueness and highest competitiveness.

In all discussed cases, information technology, process technology or materials interact intensively, a constellation that will unwrap even much higher potential in the near future e.g. by embedding more product- or processing information into the materials (e.g. factory DNA) or by engaging actuators sensible to updated information, which will demonstrate most effective local processing loops or propose modular equipment that is self configuring or self-adapting to tasks via sensor-actuator-process interactions. Information technology as well as materials and processes may be further upgraded by nanotechnology, which again catalyses more interaction potential (e.g. quality options on nano-scale, micro-fabrication of parts and pieces).

All examples show concurrent actions on many levels. The concurrency principles, especially the encapsulation, carrying self-similarity, are crucial allowing controlled changes in objectives as well as instant adaptations to new situations without recalculating all plans. Criticality thinking with its specific implementations, e.g. frozen zones or tolerance based logic via tentative variations of the space of activity or spaces of decision, plays a key role in all control set-ups. Controlled changes in objectives or resource calculations, verified by instant simulation runs, are only possible, if these are backed by continuous levelled decision logic, as proposed. In conventional planning setups, such interventions would trigger high-frequency changes, frequent new plans and bullwhipping on the process stages' demands, so online data are often ignored for protecting existing plans. The proposed setup does encourage making use of all real time data and online monitoring findings; the framework comes to life by this information and the long term effects result in higher competitiveness and better market positions.

Methodically it became clear, that the most striking effects are achieved by the interaction of different aspects, backed by own disciplines. Intensive interaction between the process layer and the informational layer e.g. combines manufacturing fields with information science pushing improvements in process qualities and efficiencies. Closed loops to economic-financial views allow us to instantly verify, to adapt and to optimise. The link to the strategy allows modifying objectives and objective bundles in situations, where rigidly imposed once well defined priorities become most unfavourable. Moreover, intensive work at intersections of fields as decision theory, social sciences and information science paves ways to new prosperous research fields, as the hybrid decision outcome proves, and this is just one outcome of many. New fields, involving augmented realities, cognition science and process or material knowledge can already be sketched; work on the edges of materials sciences, nano-systems and adaptive control, resulting in material-tool-machine real-time closed loop interactions on the base of pattern cognition and computer aided design (CAD) information, is already in implementation.

All decisions on implementation of smart units in manufacturing as well as virtual resources as services are part of a company's core strategy and cannot be delegated to IT experts or providers. The services required by manufacturing exceed general services; the challenging points for services for manufacturing, are utmost interaction capability; process steps for building value chains, powerful functionalities (manufacturers streamline business processes and optimize inventory), real-time abilities and multi corporation set ups. Distribution of manufacturing addresses main processes, hence key productivity issues. Implementations will instantly translate into lower costs; early adopters might set unseen KPI benchmarks and immediately cause competition pressure. Specialisations of manufacturers, engaging complex and expensive machinery or factories to develop certain products or sub-products for other manufacturers, are largely facilitated. Therefore, many implementations demonstrate drastic changes in the forms of manufacturing or manufactured products and initiate novel business models with new services and new products.

Networking between humans (socio informal layer), between companies, between customers and suppliers will increasingly be supported by automation, coming up to the point where human decisions are completely taken over by software. Triggered by these developments, quite a number of jobs are bound to be informatised, notably for manufacturing and manufacturing-near jobs underpinning the widespread belly feelings that the digital native will be dominating on the shopfloor at long term. The development will further challenge HR policies. If analogies to distributed automation hold, there will be mainly ICT skilled experts around in all manufacturing areas. General use of smart services should be sensitive to potential cultural and organizational differences in users' motivation to participate. Collaboration is not always considered appropriate and reasonable across cultures. Control beliefs should be managed by informing and empowering users. An intervention button or a similar type of "emergency feature" to disconnect the access of a provider on the user's smart unit is relatively easy to implement. Another major implication is that firms should train and provide general guidance to implementations, although not all are front-line employees in a traditional sense, in order to improve the social interaction skills. It is important to understand that the introduction of smart interactive services substantially changes the way providers, manufacturers and customers interact. The convergence of key information technologies is only at the beginning of an era; the first humanoid robot, able to replace humans on the shop floor, is expected to appear latest by 2025. Important studies from renowned institutions indicate rationalisation effects that could cut the workforce in industry down to 50 % within the next 10 years. The remaining half will have skills that differ from today's qualification schemes (Davis and Edgar 2011). The shopfloor will be a domain for digital experts, placing emphasis on developing IT skills and new-media literacy. Employee involvement and man machine interfaces have always been a hot research spot; functionalities and user interfaces will continue to provide a plethora of problems for intensive and sustainable however the appearance of strong body area networks (BAN) and powerful personal digital assistant devices (PDA) have already solved many of the relevant issues.

Moreover, these shifts will no longer allow ignoring basic properties of manufacturing setups and processes and their topological nature. Accepting these rules implies the unconditional priority of the process, outdating all factory centred views of manufacturing. Networkability will gain utmost importance on all levels, be it for KPIs, additionally introduced network parameters or network rules. Management could get prepared for situations where networkability and alignment parameters gain higher priority in comparison to traditional local KPIs.

Observing the players from telecommunication, hardware makers, software designers and systems integrators and the innovation power behind, it is obvious that there will be more intriguing innovations ahead. Controls of machines, robots and other equipment may be upgraded to emulate all capabilities and functions in order to ensure IP interoperability. Multi-agent systems navigate units by polling and negotiating functionalities to build up optimum process sequences. Both, product design and equipment design will have to be revised completely; increasing portions of manufacturing will become information, further optimising resources' consumption, and instigating the reuse of material as well as the after-use of products. Companies should prioritize to upgrade their equipment and to take "smart" investment decisions on new machines.

More standards on all levels will be defined, most likely on international level, and done by institutions outside of manufacturing. For implementation decisions, it's rather a matter of choosing and evaluating than developing own standards or engaging in standardisation bodies. It is always worthwhile to keep an eye on rapidly spreading devices of telecommunication and respective freeware for general use that could eventually establish irresistible quasi-, de facto or official standards. Manufacturing near associations ought to provide timely recommendations, which existing or upcoming standards should be considered.

Apart from the possible criticism for its novelty, there is an open controversy with this kind of manufacturing virtualization, in which IT has a starring role. Although recent developments and events in manufacturing virtualization give arguments to underpin IT so far, in the near future, however, it is very likely that we shall experience the follow-up rebound in organizations and on the employees' levels with striking effects there. Like a number of times before in manufacturing history, ICT just seems to make the pace. After the saturation, especially the exaggerations will strike back and will push organizations and collaborations on all levels. However, the actual virtualization wave in manufacturing is, despite of obvious gaps and high risks, unstoppable, as the advantages of these technologies are striking and their enormous benefits indispensable.

Management should be aware of alternative network configurations at any time and should have evaluations ready. Time and history will, in most cases, inhibit to instantly switch to the optimum network configurations. It will only be possible with some delay. Nevertheless all alternatives should be prepared as ready plans to be activated, as soon as implementation situations occur. Companies should continuously question their strategies; business models are jeopardised, constantly flowing, and key competencies keep repositioning. Especially, Cloud Manufacturing allows easy integration of applications and processes, both within an organization and between different organizations that wish to collaborate. However, some of the greatest concerns are security problems, loss of control (infrastructure, services, and management), technology, difficulty in migrating to other platforms, and loss of reliability. Companies therefore feel most attracted to the hybrid cloud, an option that might be reserved for applications, which do not require full synchronization or highly specialized expensive equipment. Initially, hybrid solutions with large portions of proper company implementations are expected. Eventually, competition pressure will come from these companies that are ready for taking higher risks in outsourcing ICT support, as the resulting effects are amazing. Inside and outside of companies, there will be three top priorities for information, data, and procedures: 1. Security, 2. Security, and 3. again Security!!

References

Bauer W, Schlund S, Marrenbach D, Ganschar O (2014) Industrie 4.0 – Volkswirtschaftliches Potenzial für Deutschland. BITKOM, Faunhofer IAO, Berlin, Stuttgart

Bauernhansl T (2014) Die Vierte Industrielle Revolution – Der Weg in ein wertschaffendes Produktionsparadigma. In: Bauernhansl T, ten Hompel M, Vogel-Heuser B (eds) Industrie 4.0 in Produktion, Automatisierung und Logistik. Springer Fachmedien Wiesbaden, pp 5–35

Berger R (2014) Den Vorsprung sichern. Handelsblatt (18.07.2014)

Bibel W (2011) KI ohne Geist. Künstliche Intelligenz 25(4):299–302

Braglia M, Petroni A (2000) Towards a taxonomy of search patterns of manufacturing flexibility in small and medium-sized firms. Omega 28(2):195

Burr G (2009) Führung dezentraler und teilautonomer Leistungseinheiten. In: Bullinger H-J, Spath D, Warnecke H-J, Westkämper E (eds) Handbuch Unternehmensorganisation. Strategien, Planung, Umsetzung, 3rd edn. Springer, Berlin, pp 983–1004

Busch MW (2008) Kompetenzsteuerung in Arbeits- und Innovationsteams: eine gestaltungsorientierte Analyse (1. Aufl. ed.). Gabler, Wiesbaden

Ciompi L (1997) Die emotionalen Grundlagen des Denkens: Entwurf einer fraktalen Affektlogik. Vandenhoeck & Ruprecht, Göttingen

Costanzo F, Kanda Y, Kimura T, Kuehnle H, Lisanti B, Srai JS et al (2009) Enterprise organization and operation. In: Grote K-H, Antonsson EK (eds) Springer handbook of mechanical engineering. Springer, Berlin [u.a.], pp 1267–1359

Crowcroft J, Moreton T, Pratt I, Twigg A (2004) Peer-to-peer technologies. In: Kesselman IF (ed) The grid 2, 2nd edn. Morgan Kaufmann, Burlington, pp 593–622

Davis J, Edgar T (2011) Smart manufacturing as a real-time networked information enterprise, SMLC (501c6). https://smartmanufacturingcoalition.org

DIN EN 62264-1:2014 Integration von Unternehmensführungs- und Leitsystemen - Teil 1: Modelle und Terminologie

Goll J, Dausmann M (2013) Architektur- und Entwurfsmuster der Softwaretechnik mit lauffähigen Beispielen in Java. Springer Vieweg, Wiesbaden

Kahneman D (2003) Maps of bounded rationality: psychology for behavioral economics. Am Econ Rev 93(5):1449–1475

Kilgore R, Voshell M (2014) Increasing the transparency of unmanned systems: applications of ecological interface design. In: Shumaker R, Lackey S (eds) Virtual, augmented and mixed reality. Applications of virtual and augmented reality, vol 8526. Springer International Publishing, Cham, pp 378–389

Kuehnle H (2014a) Smart units in distributed manufacturing (DM)—key properties and upcoming abilities. Paper presented at the 18th annual Cambridge international manufacturing symposium

Kuehnle H (2014b) Virtualization of production and collaborative processes in distributed manufacturing (DM) In: ICBTT 2014. Japan Society of Mechanical Engineers, pp 38–53. Congress proceedings: ICBTT2014; 7, Magdeburg, 12.04-06

Kuehnle H, Lüder A, Heinze M (2011) Criticality based decentralised decision procedures for manufacturing networks exploiting RFID and agent technology. In: Kreowski H-J, Scholz-Reiter B, Thoben K-D (eds) Dynamics in logistics. Springer, Berlin, pp 351–363

Kurzweil R (2014). Menschheit 2.0 Die Singularität naht (2, durchgesehene Aufl ed.). Lola, Berlin

Lewis K, Lange D, Gillis L (2005) Transactive memory systems, learning, and learning transfer. Organ Sci 16(6):581–598

Lödding H (2008) Verfahren der Fertigungssteuerung: Grundlagen, Beschreibung, Konfiguration (2, erw. Aufl. ed). Springer, Berlin [u.a.]

Lukas UFv, Stork A, Behr J (2014) Industrie 4.0—Evolution statt Revolution. Visual Computing beflügelt die Industrie der Zukunft. wt Werkstattstechnik online, 104(4):255–257

Owen J, Inman R, Blumenfeld D (2011) Practical modeling in automotive production. In: Kempf KG, Keskinocak P, Uzsoy R (eds) Planning production and inventories in the extended enterprise, vol 152. Springer, New York, pp 249–269

Reitmeyer T (2000) Qualität von Entscheidungsprozessen der Geschäftsleitung: eine empirische Untersuchung mittelständischer Unternehmen. DUV, Wiesbaden

Schuh G, Lenders M, Nussbaum C, Kupke D (2009) Design for changeability. In: ElMaraghy HA (ed) Changeable and reconfigurable manufacturing systems. Springer, London, pp 251–266

Simon HA (1991) Bounded rationality and organizational learning. Organ Sci 2(1):125–134

Vasseur J-P, Dunkels A (2010) Interconnecting smart objects with IP the next internet. Elsevier, Amsterdam [u.a.]

Yin JW, Zhang WY, Li Y, Chen HW (2009) A peer-to-peer-based multi-agent framework for decentralized grid workflow management in collaborative design. Int J Adv Manuf Technol 41(3/4):407–420

Conclusions and Outlook

The networkable device of the smart distributed manufacturing unit with mobile properties, working on the base of new principles, strongly informatising and decentralising manufacturing, has definitely stepped into the shopfloor to stay. Evolving network decision procedures, fully accepting the network nature of manufacturing, will replace the conventional, time slicing, command and control, one-time static, machine centred planning and control approaches as still mostly implemented in production control and enterprise resource planning. Fully incorporating network principles and, thus, making use of the network nature of manufacturing are the preconditions to overcome the particularities of high-frequency planning effects, as resource over-consumption and organisational hectic on the shopfloors. Cyber Physical Production Systems and the Internet of things challenge machines and equipment to become online, interconnected and interactive, which is fundamentally different from batch oriented logic of numerical control (NC) or Flexible Manufacturing Systems (FMS). More and better data will not cause more hectic, more frequent changes or more staccato revised decisions; accurate real-time data will rather install and maintain resource efficient, steady and smooth, easily manageable manufacturing progress, and enable to add more value in shorter time with far less input. First examples of implementing these principles already demonstrate astonishing results, and this is only the start. Witnessing the big players from telecommunication, hardware producers, software designers, and systems providers, and the huge innovation power behind, gives an impression that there will be intriguing novelties ahead in all branches. In distributed manufacturing, we are just at the beginning of an era of smart devices in all sectors; there might soon come up smart DM solutions, which we cannot even imagine today.

H. Kühnle and G. Bitsch, *Foundations & Principles of Distributed Manufacturing*,
Springer Series in Advanced Manufacturing, DOI 10.1007/978-3-319-18078-6

Which Essentials Does the Book Cover?

With comprehensive description of this new working field, the book proposes solid ground to distributed manufacturing as a newly emerging discipline. This discipline will certainly live from outside contributions, theories from other areas and accentuated impacts of networking. Irrefutable evidence and motivation for interdisciplinary interaction are elaborated in Chap. 3, where the backgrounds of philosophy of sciences give important sidelines for the design of frameworks. Strong interactions between disciplines are instigated, not just limited to borrowing from outside but to changing and extending the core of the manufacturing field. By the topological construct of the Hausdorff space, a comprehensive and powerful frame is suggested as a base. This frame has already been an important tool for substantial work since 10 years, initially for fractalising factory network set-ups and for advanced models and procedures in collaborative facilities planning and process implementations; actually, this approach gets full support by the most recent developments around the Internet of Things and the Cyber Physical Systems in general as well as in the context of Distributed Manufacturing.

As the core of the set-up, a generic construct is proposed, using six layers, which catalyses interactions and interchanges between disciplines and, at the same time, preserves disciplines' aspects throughout the entire network on all levels of detail. Important attached models are introduced, especially in the demand, resource and decision-making context, and tied together by a decision space model, fully incorporating the criticality thinking from (neuro-)network sciences into manufacturing. Moreover, complexity, criticality and simultaneity establish manufacturing specific concurrency principles, synthesising the network levels of detail, and decision modes with network and process configuration issues. Key portions are immediate outcomes of applications of topological spaces, their fractal dimensions along with their patterns of self similarity. At this point, the full embedding of the established model worlds from manufacturing and manufacturing management, already known as the bases of enterprise software solutions, smoothly bridges from conventional manufacturing set-ups to the world of smart Distributed Manufacturing. As decision structures have to be considered key components in any manufacturing setup, an upcoming way of hybrid decision making is outlined in detail. Manufacturing units will exhibit additional properties to become part of smart Distributed Manufacturing networks in the defined sense. To meet the evolving character of these properties and their smartness, a Distributed Manufacturing Maturity Model (D3M) is proposed for evaluating technological readiness levels and smartness' achievements.

All instruments, set ups and models are embedded into significant and profoundly discussed Distributed Manufacturing implementation examples. The value of these examples in Chap. 6 is threefold, as they provide for:

- Verifications of all given theory portions and generic models, including the concurrency principles and the embedding of conventional model worlds

- Solid demonstrations of the suggested network decision mode as being iterative, gradual and levelled, exhibiting criticality thinking and fold/unfold mechanisms
- Irrefutable testimonials for the enormous power of smart distributed manufacturing implementations, underpinned by hard fact evidence of shifts in KPIs and future improvement potential.

Moreover, developments in further converging technology fields towards distributed manufacturing are anticipated in this outlook chapter.

What is the Value of the Book?

The book proposes the first comprehensive theory framework that treats manufacturing as distributed networks. The topological mapping, as introduced, easily succeeds in including the network nature of manufacturing processes from the basics. Less functional, less restricted, organisationally lower bounded and free from habitual domains, the approach also deeply incorporates a number of novel network possibilities into Distributed Manufacturing management and control. Abilities for comparing plans to real-time monitored process states unearths so far unseen potential for optimising resources' input and process efficiencies. The proposed decision cycle, the use of the concurrency principles, and the criticality thinking with the toolboxes offered allow the implementation of new procedures for control and management, which prove to more stable, come closer to the real events and which are, at the same time, simpler and more effective. Thereby, easy ways towards the next generation network resource planning are demonstrated for making full use of the smart units' properties and the Real Time Enterprise (RTE).

Manufacturing equipment producers and machine toolmakers find a comprehensive picture for mirroring their reconsiderations of design activities; they may take out orientation on the consequences, the risks, the market opportunities and innovation options for value creation with all involved objects co-operating online and interconnected. On top of all the excellent features and capabilities, modern manufacturing units and production equipment exhibits already, the units will unlock a variety of helpful IoT services that are increasingly promoted by external service providers. Continuous rethinking of all value chains and organisational set ups will become irrefutable permanent tasks for company planners and strategists. They will find both, profound discussions of the upcoming properties of manufacturing equipment, and demonstrations of the game changing rules in manufacturing networks. Company innovators will find valuable hints and proposals to verify their work and for taking their next steps on more stable decision grounds.

For shopfloor practitioners and system developers, sound descriptions of application examples are given. These examples are deliberately taken from a variety of distributed manufacturing implementations, displaying cloud computing, item localisation and ordertracking in chemical industries on one side, and smart extensions of advanced implementations in metal processing and auto supplier industry on the other.

Researchers find consolidated ground for further work in many directions; solid scientific base grants well-staked research fields; the roots of manufacturing are fully kept in mind while the network nature of manufacturing processes is acknowledged, an inevitable move for completely exploiting the Distributed Manufacturing's smart options. A framework for the field of Distributed Manufacturing including the interdisciplinary contributions for promising theory building is given as starting point for further research. The Hausdorff space, much earlier proposed and verified as suitable formal model for advanced manufacturing concepts by the authors already, obviously finds its practical translations in these days, e.g. as the cyber physical production system or smart object. The mappings, exhibited there, are technically realised by virtualisations of objects and units e.g. implemented as virtual machines (VM) into manufacturing equipment. Moreover, the framework provides adequate instruments for capturing complexity and variety, as desired. Fast progress in fragmentations and atomization of process steps and miniaturisation of technologies urged for more comprehensive, formal descriptions of units and process fragments, and better design to hold the networks together. The compounds are simple and, put together in the right way, at the same time, rich enough to cover all relevant manufacturing aspects in their full ranges.

For scholars and academics, the value of the book is manifold. There are sections, treating structuring and realigning manufacturing networks on one hand, and sections for reconsidering manufacturing units' smartness and decision models for planning, control and optimisation on the other. Factory centred thinking will be replaced; network guided views change all well-known models' roles. For the first time, a coherent approach for unfolding manufacturing networks into levels of details is proposed. This is only possible by reclassifying models into adequate sets, as visualised, and their consolidation according to standardised levels of detail. Moreover, with the Distributed Manufacturing Maturity Model, another visualising tool is made available.

Lecturers find a framework that incorporates complexity thinking, network view, structures to capture contributions from outside disciplines, and a setup for network management and control into manufacturing on coherent theory base. The theory set up is capable to do both, embedding conventional manufacturing solution spaces with the models used there, and flooring the new fields coming in with the novel devices and progresses in information technology in general. As a number of disciplines are involved, clusters of preferred disciplines are suggested for fertile integration in the case example discussions. For encouraging and facilitating interdisciplinary study work, a catchy layer construct is proposed. Pathways are paved for instigated interdisciplinary interactions right from the core, able to speed up both, project implementations and study work. Moreover, exhaustive examples, unifying the established models with novel procedure for manufacturing network evolution, are detailed and discussed. The book is kept free from formalisms and mathematical notation to make it accessible for practitioners and manufacturing students. The outline smoothly bridges existent solutions and the rapidly spreading new thinking. It also gives valuable support for training staffs in specialised manufacturing areas and grants easy and structured access to this rapidly emerging manufacturing network world.

Looking Ahead!

From what is already known about networking of enterprises will now be observed on lower levels of detail as well. Networked machines, equipment and parts will even be found on the field control level in the context of flexible manufacturing units and material. Collaboration of cyber existence, Internet of things and World Wide Web may certainly be addressed as the key enablers to shape real time enterprises. Pervasive computing has been one of the triggers of distributed manufacturing in just some areas. There is certainly much more to come, especially if the imagine embedded processors in practically every object and equipment. Mechanisms can be imagined for virtually composing products of intelligent components finding each other on the path of value creation; this option has not been addressed in advanced manufacturing concepts so far.

The huge potential of Distributed Manufacturing appears evident in a single company or factory already, whereas it is the option for multi site and multinational value and supply chains with most gains. Highly flexible and versatile structures may be exploited in global dimension most effectively; following this projection, entire manufacturing sectors may soon come up with overall resource sharing implementations, driven by remote cloud manufacturing (RCM) as well as higher integration of the product design process into Distributed Manufacturing, the Cloud-Based Design and Manufacturing (CBDM).

Another strongly upcoming field will be empowered machine to machine M2M communication; novel interaction modes concerning person to machine P2M are on the way. Smart technologies provide collaborative environments promoting adaptive agility, transparency and empowerment through more effectiveness. These environments easily outpower conventional setups, since everything is mobile and wireless and literally every feature is topped at scale. Waiting with implementation means obsolescence and decaying competitiveness!

Increasing portions of manufacturing will become information; optimisation of resource consumption will instigate the reuse of materials and promote the after-use of products. The term Distributed Manufacturing comprises the distribution of integrated resources as well as the integration of distributed resources; encapsulations of resources, units and object's models may be increasingly seen as parts of the cloud. All companies are facing ever-growing data volumes and computer power requirements, therefore cloud computing and big data shake up the service- and the outsourcing business; third parties will offer "Anything as a Service" (AaaS) as commodities on demand, scalable and always on the latest standards. This will again make manufacturing companies more dependent on IT providers. After visible reluctance of the leading software vendors to deal with novel technologies, meanwhile all cutting-edge players are fully involved leaving few alternatives to the competitors.

Wireless technologies have also brought direct telecommunications' involvement into manufacturing; this involvement is strongly gaining ground e.g. by efficient tracking systems, GPS positioning, smart metering, and wireless body area networks

(WBAN). Manufacturers of computer hardware as well as software vendors will have to take into account these developments and the revolutionising consequences thereof.

By definition, we include additive manufacturing to the context of Distributed Manufacturing, without further mentioning solutions or emphasising this technology in theory and examples. As far as the theory is concerned, additive manufacturing units may be smart units as any other unit, as described. Standard triangulation language STL may be emulated as any other control compound. The attention, additive manufacturing receives in recent studies, will certainly accelerate developments in the direction of shortening or eliminating process chains by more condensed information. Smart materials or materials with so far unseen attributes will open new horizons as e.g. zirconium ceramics or normal fibre materials. This is no contradiction, rather the full endorsement to our approach, magnifying the scope of Distributed Manufacturing by more options of design anywhere manufacture anywhere (DAMA). One of the key drivers, here again, is the Internet of things with the globally networked smart manufacturing unit.

We often discuss questions around what the future might bring for Distributed Manufacturing and smart automation; smartness of devices and its online awareness trigger peoples' imagination and fantasy, instantly evoking ideas and trajectories that could grasp future developments. Most experts agree that Moore's Law will still be valid for some time, so micro devices will gain enormous capabilities and will strongly promote progressing modularity of equipment and machines, so e.g. robots could adapt to literally every task by orchestrasting directly interacting smart modules. A crucial precondition is powerful M2M communication, which is broadly on the way. Furthermore, the rising impact of artificial intelligence (AI) is often pointed at. It is exactly here, where we see the highest potential, as the outdated top-down approach for modelling brain functions is being substituted by decentralised bottom-up mechanisms based on networks. Hebb's rule, highlighting "trial, error and incentive", may directly be applied as outcome of the concurrency principles in Chap. 4. The criticality thinking, introduced here, originates from network sciences and, indeed, neuro-scientific models of the human brain. Times are near, where smart machines will be more apt to do manufacturing tasks than humans will, at least for certain things, but specific abilities will surely remain genuine human domains. Rather than man versus machine scenarios, we see the smooth symbiotic mergence of the two, for highly sophisticated tasks' execution; a first important step in this direction is made by fully incorporating hybrid decision-making; here, we see a growing field that further fertilises all upcoming distributed manufacturing developments.

Both, product designs and equipment designs will rapidly progress. In the convergence of different technologies and disciplines we are only at the beginning of an era. Even though, the addressed technologies plus surely additional ones, merging nanotechnology and cognition sciences, exponentiating by information networks will surely bring about many more intriguing innovations, unexpected additional applications and astonishing novel solutions in manufacturing; the proposed theory approach is supposed to be ample enough to capture them all.

With this book, the proposed framework, the outlined theory approach with verifying instructive case studies, we strongly feel to step into the right direction and to speed up further developments on this fascinating research field of Distributed Manufacturing.

Annex

Fractals and Self Similarity

An object is considered to be self-similar if it looks about the same on any scale. **Fractals** are a particularly interesting class of self-similar objects. Self-similar objects with parameters described by a power law, such as $N = S^d$ where $d = \frac{\ln N}{\ln s}$ is the dimension of the scaling, are called fractals. It is also known as the Hausdorff dimension. Fractals play a decisive role in mastering complexity, as applied in organisation and factory operations as well. This view enables the introduction of self organisation, self optimisation and self structuring into or factory organisations. As this outline proves, the view can be extended into decision and aspect constructs.

References

Kühnle H (2005) Fractals extended enterprise: framework and examples for multi-party supply chains. In: Shuping Yl (Hrsg.), Chen X (Hrsg.), Yang Y (Hrsg.) (eds) Modern industrial engineering and innovation in enterprise management, IEEM 2005, 12th international conference on industrial engineering and engineering management, Chongqing, China, proceedings, vol 1, pp 211–217. China Machine Press, Beijing, 6–8 Nov 2005

Kühnle H (2009) Self-similarity and criticality in dispersed manufacturing—a contribution to production networks control. In: Dispersed manufacturing networks, pp 59–76. Springer, London, ISBN 978-1-8488-2467-6

H. Kühnle and G. Bitsch, *Foundations & Principles of Distributed Manufacturing*, Springer Series in Advanced Manufacturing, DOI 10.1007/978-3-319-18078-6

Topological Hausdorff Space

Be X an arbitrary set. As set of subsets of $X \in T$ is named a Topology on X, if:
(1) $X \in T$
(2) every joint of the set of T is included in T
(3) finite intersections of sets of T are included in T,
so the pair (X; T) is then called **topological space**, the elements of T are called open sets of X.

The definition of topological spaces is very general and multiply usable, especially for describing rich concepts of spaces. Further specifications are very much compatible with our spatial intuitions in the context of work Euclidean spaces. Therefore, another definition may be given to obtain more appropriate spaces for distributed manufacturing considerations.

A topological space X is called **Hausdorff** if every pair of points can be separated by open sets.

That is, if $x_1 \neq x_2 \in X$ then there are disjoint open sets U_1 and U_2 with $x_1 \in U_1$ and $x_2 \in U_2$.

The idea behind the definition of the original space is to offer mappings that include properties of continuity, so the space elements and their later mappings are closely related.

If X and Y are topological spaces, a (homeo)morphism from X to Y, Ø is a map X to Y, so Ø and their interactions are continuous. This definition leads to the conservation of properties of the spaces when projecting onto other spaces.

A topological space X is said to be a **Hausdorff space** if given any pair of distinct points, there exist neighbourhoods; this property is often summarized by saying "points can be separated by open subsets". The Hausdorff property ensures that a topological space has subsets, all to form to our spatial intuition. For many purposes, it is useful to restrict attention to spaces that do not have too rich subsets. This construction may be ideally used for attaching spaces that represent reductions of the original, as valuable for CPS.

On these open subsets topologies can be defined that are related with the topologies and are, under certain conditions, called **quotient topologies** induced by maps **f**. The theory of quotient spaces gives ways to construct new topological spaces by attaching. The quotient spaces Y are also called **adjunction spaces**, and is said to be formed by **attaching to** X along **f**. The map f is called the attaching map. This view captures the intuitive idea of attached spaces e.g. alike Euclidean spaces, spaces and we daily deal with. For virtualisation is of objects, and this view gives the sidelines for all constructs use in distributed manufacturing, as not the systems itself but models thereof are communicating, interacting and linking. As several aspects are involved that underpins the importance of different mappings originating from the same unit and still being somehow related. Negation interaction and linking can be done with any of the mappings without contradictions, as long as correspondent granularity is ensured. This gives also sidelines for standardisation needs in distributed manufacturing.

References

Lee JM (2013) Introduction to smooth manifolds, 2nd edn. Springer, New York, ISBN 978-1-4419-9981-8
Querenburg Bv (2001) Mengentheoretische Topologie. Springer, Berlin, ISBN 3540677909

Bibliography

Adelstein F, Gupta SKS, Richard GG et al (2005) Fundamentals of mobile and pervasive computing. McGraw-Hill, New Delhi

Balakirsky S (2015) Ontology based action planning and verification for agile manufacturing. Robot Comput Integr Manuf 33:21–28

Bennet A, Bennet D (2004) The intelligent complex adaptive system. KMCI Press Elsevier, Amsterdam

Borangiu T, Thomas A, Trentesaux D (eds) (2015) Service orientation in holonic and multi-agent manufacturing, vol 594. Springer, Berlin

Brecher C (ed) (2015) Advances in production technology. Springer, Switzerland

Camarinha-Matos LM, Afsarmanesh H (2008) Collaborative networks: reference modeling. Springer, Heidelberg

Davis E, Spekman R (2004) The extended enterprise: gaining competitive advantage through collaborative supply chains. FT Prentice Hall, New York

Dekkers R (2013) Applied systems theory. Springer, London

Dey AK, Abowd GD (2000) Towards a better understanding of context and context-awareness. In: Proceedings of the workshop on the what, who, where, when and how of context-awareness. ACM Press, New York

Glaser BG, Strauss AL (1967) The discovery of grounded theory: strategies for qualitative research. Industrial Press—Technology & Engineering, Aldine, New York

Kaku M (2011) Physics of the future: how science will shape human destiny and our daily lives by the year 2100. Penguin, Auckland

Kilgore R, Voshell M (2014) Increasing the transparency of unmanned systems: applications of ecological interface design. In: Shumaker R, Lackey S (eds) Virtual, augmented and mixed reality. Applications of virtual and augmented reality, vol 8526, pp 378–389. Springer, New York

Krar SF, Gill A (2003) Exploring advanced manufacturing technologies. Industrial Press, New York

Kuehnle H (ed) (2010) Distributed manufacturing—paradigms, concepts, solutions and examples. Springer, London

Kuhn T (1962) The structure of scientific revolutions. University of Chicago Press, Chicago

Li W, Mehnen J (eds) (2013) Cloud based design and manufacturing: distributed computing technologies for global and sustainable manufacturing. Springer, London

Liu N, Li X, Shen W (2014) Multi-granularity resource virtualization and sharing strategies in cloud manufacturing. J Netw Comput Appl 46:72–82

Lu Y, Xu X, Xu J (2014) Development of a hybrid manufacturing cloud. J Manuf Syst 33(4):551–566

H. Kühnle and G. Bitsch, *Foundations & Principles of Distributed Manufacturing*,
Springer Series in Advanced Manufacturing, DOI 10.1007/978-3-319-18078-6

Luhmann N (1990) Die Wissenschaft der Gesellschaft. Suhrkamp, Frankfurt

Metz D (2014) The concept of a real-time enterprise in manufacturing. design and implementation of a framework based on EDA and CEP. Springer Fachmedien Wiesbaden, Wiesbaden

Mintzberg H (2005) Developing theory about the development of theory. Oxford University Press, Oxford

Oesterle H, Fleisch E, Alt R (2000) Business networking, shaping collaboration between enterprises, 2nd edn. Springer, Berlin

Pal K, Karakostas B (2014) A multi agent-based service framework for supply chain management. Procedia Comput Sci 32:53–60

Popper K (1959) Logic of scientific discovery. Hutchinson: London

Querenburg Bv (2001) Mengentheoretische Topologie. Springer, Heidelberg

Schaefer D (ed) (2014) Cloud-based design and manufacturing (CBDM) a service-oriented product development paradigm for the 21st century. Springer, Cham

Vasseur JP, Dunkels A (2010) Interconnecting smart objects with IP: the next internet. Morgan Kaufmann, New York

Index

H. Kühnle and G. Bitsch, *Foundations & Principles of Distributed Manufacturing*,
Springer Series in Advanced Manufacturing, DOI 10.1007/978-3-319-18078-6

MIX
Papier aus verantwortungsvollen Quellen
Paper from responsible sources
FSC® C105338

If you have any concerns about our products,
you can contact us on
ProductSafety@springernature.com

In case Publisher is established outside the EU,
the EU authorized representative is:
Springer Nature Customer Service Center GmbH
Europaplatz 3, 69115 Heidelberg, Germany

Printed by Libri Plureos GmbH
in Hamburg, Germany